Shifting Frontiers of *Theobroma Cacao* - Opportunities and Challenges for Production

Edited by Samuel Ohikhena Agele and Olufemi Samuel Ibiremo

Published in London, United Kingdom

Shifting Frontiers of *Theobroma Cacao* - Opportunities and Challenges for Production
http://dx.doi.org/10.5772/intechopen.107709
Edited by Samuel Ohikhena Agele and Olufemi Samuel Ibiremo

Contributors
Alex Asante Appiah, Julia Bello-Bravo, Anne Namatsi Lutomia, Araz Meilin, Endrizal, Busyra B. Saidi, Jumakir Jumakir, Waluyo, Yardha, Muh. Asaad, Syafri Edi, Yustisia, Sigid Handoko, Nila Wardani, Suparwoto, Handoko, Nurmili Yuliani, Nurhayati, Herma Cahyaningrum, Rein E. Senewe, Saidah Saidah, Herlina Nanny Salamba, Delima Napitupulu, Ismon Lenin, Risma Fira Suneth, Nur Imdah Minsyah, Julistia Bobihoe, Antoine Alban Kacou M'bo, Mahyao Germain Adolphe, Kouakou Kouadio, Mamadou Cherif, Renée Brunelle, Alla Kouadio Okou, Daouda Koné, Adama Bamba, Evelyne N'Datchoh Toure, Yanick Rouseau, Samuel Ohikhena Agele, Olufemi Samuel Ibiremo, Oladitan Titilayo, Temitope S. Egbebiyi, Christopher Lennard, Pinto Izidine, Romaric C. Odoulami, Piotr Wiolski, Akintunde I. Makinde, Kurnia Dewi Sasmita, Nur Kholis Firdaus, Edi Wardiana, Dibyo Pranowo, Iing Sobari, Saefudin, Nur Kholilatul Izzah, Asif Aunillah, Sakiroh, Maman Herman, Dewi Listyati, Sampson Addae, Sarah Acquah, Samuel Nyarko Essuman, Hélio Rocha Sousa Filho, Raildo Mota de Jesus, Marcos de Almeida Bezerra, Jorge Chiapetti, Julian Villa-Turek Arbelaez

First published in London, United Kingdom, 2024 by IntechOpen
IntechOpen is the global imprint of INTECHOPEN LIMITED, registered in England and Wales, registration number: 11086078, 5 Princes Gate Court, London, SW7 2QJ, United Kingdom

British Library Cataloguing-in-Publication Data
A catalogue record for this book is available from the British Library

Additional hard and PDF copies can be obtained from orders@intechopen.com

Shifting Frontiers of *Theobroma Cacao* - Opportunities and Challenges for Production
Edited by Samuel Ohikhena Agele and Olufemi Samuel Ibiremo
p. cm.
Print ISBN 978-1-83768-319-2
Online ISBN 978-1-83768-320-8
eBook (PDF) ISBN 978-1-83768-321-5

For EU product safety concerns:
IN TECH d.o.o., Prolaz Marije Krucifikse Kozulić 3, 51000 Rijeka, Croatia,
info@intechopen.com or visit our website at intechopen.com.

Meet the editors

Samuel Ohikhena Agele is a professor and lead researcher in the Plant Physiology and Ecology Group, Department of Crop, Soil and Pest Management, Federal University of Technology, Akure, Nigeria. His research interests include plant physiology and ecology, agricultural climatology, agricultural water management, conservation agriculture, food security, bioreclamation, and plant domestication. His work is at the frontiers of agriculture and the environment and his major contributions to science are in the study of agriculture and interface with global change. He has authored more than 100 peer-reviewed journal articles and has engaged in debates concerning how more severe droughts, as a consequence of climate change, can accelerate tree mortality.

Dr. Olufemi Samuel Ibiremo is the Director of Training, Cocoa Research Institute of Nigeria, Ibadan, Nigeria. His research interest is soil fertility using organic fertilizers, rock phosphates, and biotic agents (as alternative sources of nutrients) for the cultivation of cocoa, coffee, cashew kola, and tea (Cocoa Research Institute of Nigeria [CRIN]-mandated crops). He has authored more than forty peer-reviewed journal articles. He has also given talks and presented his research findings at regional and international scientific conferences, seminars, workshops, and symposia.

Contents

Preface **IX**

Section 1
Climate Change as a Threat to Growing Area Suitability for Cocoa **1**

Chapter 1 **3**
Perspective Chapter: Microclimate, Plant Stress and Extension of Cacao Frontiers to Marginal Agroecologies of the Rainforest Tropics
by Samuel Ohikhena Agele, Olufemi Samuel Ibiremo and Oladitan Titilayo

Chapter 2 **21**
Potential Impact of Stratospheric Aerosol Geoengineering on Cocoa Suitability in Nigeria
by Temitope S. Egbebiyi, Christopher Lennard, Pinto Izidine, Romaric C. Odoulami, Piotr Wiolski and Akintunde I. Makinde

Chapter 3 **39**
Climate Variability and Outlook of Cocoa Production in Côte D'ivoire under Future Climate
by Antoine Alban Kacou M'bo, Mamadou Cherif, Kouakou Kouadio, Mahyao Germain Adolphe, Adama Bamba, Evelyne N'Datchoh Toure, Alla Kouadio Okou, Renée Brunelle, Yanick Rouseau and Daouda Koné

Chapter 4 **79**
Cocoa Production and Distribution in Bahia (Brazil) after the Witch's Broom
by Hélio Rocha Sousa Filho, Marcos de Almeida Bezerra, Raildo Mota de Jesus and Jorge Chiapetti

Chapter 5 **95**
Challenges and Opportunities for Indonesian Cocoa Development in the Era of Climate Change
by Kurnia Dewi Sasmita, Edi Wardiana, Saefudin, Dibyo Pranowo, Asif Aunillah, Nur Kholilatul Izzah, Maman Herman, Nur Kholis Firdaus, Iing Sobari, Sakiroh and Dewi Listyati

Section 2
ICT Role in Advancing Climate Change Adaptation and Climate Mitigation **115**

Chapter 6 **117**
Using Information Technologies (ICTs) to Improve *Theobroma cocoa* Extension Service: Lessons from the Case of Ghanaian Bean Farmers
by Julia Bello-Bravo and Anne Namatsi Lutomia

Section 3
Climate Change, Crop Protection and the Environment **133**

Chapter 7 **135**
Evaluation of Progress in Cocoa Crop Protection and Management
by Alex Asante Appiah

Chapter 8 **157**
The Integrated Pest Management Implementation of the Cocoa Pod Borer in Indonesia
by Araz Meilin, Nurmili Yuliani, Nurhayati, Hermawati Cahyaningrum, Rein E. Senewe, Saidah, Herlina N. Salamba, Delima Napitupulu, Ismon Lenin, Risma Fira Suneth, Nur Imdah Minsyah, Handoko, Suparwoto, Endrizal, Busyra B. Saidi, Jumakir, Waluyo, Yardha, Muh. Asaad, Syafri Edi, Yustisia, Sigid Handoko, Nila Wardani and Julistia Bobihoe

Chapter 9 **173**
Impact of Illegal Mining Activities on Cocoa Pollinator Abundance in Ghana
by Sampson Addae, Sarah Acquah and Samuel Nyarko Essuman

Section 4
Cocoa Cultivation: Role in Peace Building **191**

Chapter 10 **193**
Cacao: A Path of Everyday Resistance and the Pursuit of Peace
by Julian Villa-Turek Arbelaez

Preface

This book, "Shifting Frontiers of *Theobroma cacao* – Opportunities and Challenges for Production", presents a comprehensive perspective that acknowledges the interactions of changing environmental conditions (climate, soil, and agroecological), cocoa production, and sustainability.

The book is short, clear, readable, and accessible, and covers a broad range of topics to quicken the pace of understanding and knowledge (scholarship) among agroclimatologists, social and natural scientists, policymakers, and activists concerned with burning issues of changing environment conditions and their current and future implications for cocoa production and sustainability.

The chapters are organized into four sections: "Climate Change as a Threat to Growing Area Suitability for Cocoa", "ICT Role in Advancing Climate Change Adaptation and Climate Mitigation", "Climate Change, Crop Protection and the Environment", and "Cocoa Cultivation: Role in Peace Building". The chapters inform, stimulate, and provoke thought as well as actions and partnerships relevant to addressing challenges and constraints to cocoa production in the era of a changing climate. It provides insights into the need for cocoa actors within the cocoa sector to strengthen climate mitigation and resilience building and to come to grips with the realities, magnitude, and inevitable persistence of climate challenge to cocoa production and sustainability. The authors synthesize and deploy current scholarship on climate change and its implications and ramifications for the crop *Theobroma* in ways to promote and activate intelligent thought, collective actions, and partnerships for cocoa production sustainability. The book is invaluable to scientists, researchers, agriculturists (cocoa farmers), and students, as it illuminates the challenges presented by climate change for cocoa production and sustainability.

As the editors of this book volume, we would like to express my profound thankfulness and gratitude to all the authors, reviewers, and staff at IntechOpen. We all are grateful and appreciative of their work and ideas.

Samuel Ohikhena Agele
The Federal University of Technology,
Akure, Nigeria

Olufemi Samuel Ibiremo, Ph.D.
Cocoa Research Institute of Nigeria,
Ibadan, Nigeria

Section 1

Climate Change as a Threat to Growing Area Suitability for Cocoa

Chapter 1

Perspective Chapter: Microclimate, Plant Stress and Extension of Cacao Frontiers to Marginal Agroecologies of the Rainforest Tropics

Samuel Ohikhena Agele, Olufemi Samuel Ibiremo and Oladitan Titilayo

Abstract

Cacao (*Theobroma cacao* L.) is an important cash crop of the rainforest tropics where it is a major foreign exchange earner, industrial raw material, livelihood, and offer ecosystem services. The rainforest of the tropics is noted for high productivity potential for cacao, however, and its development prospects is beset with numerous challenges among which is the threat of climate change which is setting new ecological boundaries. The new regimes of climate are expected to affect the area suitable for agriculture, thus, crop species are bound to grow in areas where they were not previously grown and areas that were hitherto not suitable for their production. Nevertheless, the shifting environment conditions and associated marginal growing environment conditions (weather: (increasing warming and drought) and soil) may offer opportunities for extending crop frontiers beyond its current ecological boundaries. It is therefore necessary to develop strategies for alleviating constraints imposed by changing environmental conditions thus setting the agenda for climate smart adaptable and sustainable production systems. In addition, efforts to unlock the potentials of the new environmental boundaries for crops will benefit from knowledge, technologies and innovations and climate mitigation.

Keywords: Theobroma, cultivation, marginal, agroecology, mitigation resilience, climate stress

1. Introduction

The global environment change (GEC) including climate change has resulted in major shifts of crop growing environments (agroecologies, weather, soil) and thus changes in the suitability of agroecologies for crops especially fruit tree species (cocoa, citrus, oil palm). Climate change has been projected to cause shifts in the present day growing conditions and set new environment boundaries and habitat range for crops and livestock, and such would crop yields and crop growing area suitability. In particular, increases in temperatures and declining rainfall had led to large parts of sub-Saharan The shifting climate in Sub-Saharan Africa has produced changes in rainfall pattern, temperature, season length and abiotic stress and shifts in suitability of crop growing areas especially for fruit tree species (cacao, oil palm, citrus, coffee etc.) along ecological transect of wet to dry forests. There is however, increasing need to extend production and thus, frontier of fruit tree cultivation to meet rising global demand along other benefits (food security, livelihood, ecosystem services). The new climatic regimes within agroecologies of Saharan Africa (SSA) will affect water resources, agriculture, food and nutrition security, livelihoods and economic growth.

Cacao (*Theobroma cacao*) is an important cash crop in Africa, Central and South America and Asia with an estimated annual production of 4.92 million tons [1]. In these regions, cacao is a major foreign exchange earner, and important as industrial raw material, livelihood, and ecosystem services provision. Based on its global importance, is the need for increased yields and expansion of cultivation to meet rising global demand for cacao beans.

In West Africa, the rainforest belt constitutes the cocoa growing regions where annual rainfall is high (ranging between 1500 and 2000 mm in a bi-modal distribution pattern, and wet-dry season transitions. The dry season is a terminal drought situation for crops especially for plantation (permanent) crops which undergo moderate to severe hydrothermal stresses [2–4].

Rainfall is most important determinant of the success of farming operation within the rainfed farming system where crops are seldom irrigated despite the climate stress of the seasonal wet and dry transitions.

In the West African cocoa producing regions, climate variability and extreme weather events constitute the greatest production constraint to crop productivity [5–7]. Climatic variability has produced significant reductions in cocoa yields and deterioration in bean quality [5, 6]. Agrometeorological studies would contribute to improved understanding of the effects of climate factors on plant environment. The understanding will help to improve farming practices, but also to assess certain risks to the cocoa tree due to irregular rainfall, extreme temperatures and humidity, and low insolation.

In the West Africa, farmers often grow crops in mixed stands (intercropping and agroforestry practices) which subject them to varying intensities of shading. Reports form cacao-based agroforestry system of tropical rainforest belt have shown that over 100 species covering 30 families of shade trees are found in cacao agroforests. Timber and fruit tree species are widely used by farmers to provide shade for transplanted cacao seedlings on the field [8, 9]. The heterogenous shading regime sin cacao agroforests are constituted by shade tree species; timber and fruit trees forming structurally complex closed canopy multi-strata system (under-growths, middle and upper storey species) which retain attributes of natural rainforest ecosystem. The agroforestry system retain the positive attributes of natural forest climax. The co-occurring species crops/trees therefore grow under variable shade intensities (20 to 70% reduction in incoming radiation) and some pockets of unshaded area (full sunlight). Thus, the variable shading scenarios in natural cacao based agroforestry systems of the humid tropics can be constituted by: A: 30:70 timber trees and fruit tree species in mixture - PAR interception ranging from 70 to 20%, B: 50:50 timber trees and fruit tree species in mixture - PAR interception ranging from 70 to 20%, and C: open field (full sun) cacao where light regimes is greater than 70% PAR interception [10].

Shade is known to buffer the effects of stress factors (high irradiance, temperatures and atmospheric deficits (low humidity) of the dry season on tropical perennial fruit tree species in plantations and lower incidence of shoot/branch die-back and tree mortality [11].

However, despite shade provision for transplanted cacao seedlings, high mortality percentages are recorded between first and second dry season. Such seedling death is attributable to high hydrothermal stresses of the terminal drought condition of the dry season [9].

The high seedling mortality constitutes constraints to establishment of seedlings in cocoa rehabilitation efforts following dry season drought. Other factors including low soil fertility, pests (termites and capsids) and poor access to improved planting materials [12].

The naturally occurring shading regimes and variable light intensities are known to expose cacao to modified microclimate, and altered availability and balance of resource use to crops (enhanced competition). Cacao-based agroforestry system ameliorated weather conditions, improved ecological benefits and ecosystem health, the resultant enhance carbon sequestration will reduce global warming.

Cacao-based agroforestry ecosystem contributes to terrestrial carbon budget via carbon storage in soil and crops, as well as microclimate amelioration, improved ecological benefits and ecosystem health. These services exert feedback on terrestrial climate. Reports indicate that the high potentials of net gains in C sequestration from cacao-based agroforestry system is a promising CO_2 mitigation strategy [11].

2. The global economic and social impact of cocoa

The cocoa industry has made enormous contribution to global socio-economies. Global cocoa production and expansion of the chocolate industry had retail value of USD 100 billion in 2021, this is expected to grow further until 2027 [13]. Other reports projected expansion growth of about USD 69.1 billion by 2030 [14, 15]. However, over 79% total annual cocoa bean production is processed into cocoa butter, chocolate, or other products outside the main producing centres especially in the Sub-Saharan Africa for example, Europe followed by Asia and Oceania [13]. Cocoa is vital for Nigeria's economy. More than 70% of cocoa farmers are smallholders with average farm size less than 1.5 ha, yield increases are mostly due to expansion of cocoa land area Cocoa yields are principally affected by soil and climatic variables [10, 16, 17]. There has been declining cocoa export from the mid-1980s to till date.

3. Agricultural ecology of the rainforest of Nigeria

The rainforest zone of Nigeria is characterized by annual rainfall of over 1500 mm distributed in a bimodal pattern of seven to 8 months duration (the growing season), and 3 to 4 months of dry season (typically from late November to march) characterized by low humidity, temperatures over 32°C and clear skies. Rainfall is characterized by unpredictable onset and cessation dates, variable lengths of growing seasons, increasing intensities and duration of terminal drought situations and elevated soil and air temperatures.

Seasonality of sowing implies that crop production is rainfed and linked to seasonally available soil water (rainfed agriculture). The four to 6 months of dry weather results in soil water deficits and high soil and air temperatures [8, 10]. In fruit trees, unfavorable e soil and weather conditions of the short and long dry seasons are known to affect flowering and pod production with negative consequences on bean size. In fruiting trees, water deficits result in lower yields and an increase in the level of mirid (capsid) damage and tree mortality. Since cacao is seldom irrigated, trees undergo severe water stress during the dry season, leading to branch die-back, and for pod-bearing trees, hydrothermal stresses affect pod filling and bean size. Additionally, the wet-dry seasonal transitions cause seasonal leaf shedding at the onset of dry season and leaflessness until beginning of rain. Under the episodes of drought of the dry season, tree crops including cacao, may be subjected to water stress-induced hydraulic failure and cavitation events in the soil-canopy continuum. Thus, fruit trees would have to adjust water relations and water use and other physiological functions for optimum productivity.

4. Climate change and extreme weather events and the changing cocoa production landscapes in West Africa

The scenarios of climate change especially, temperature and rainfall patterns for West Africa including Nigeria have been variously modeled using process-based methods of the General Circulation Models (GCM) and Simple Climate Models (SCM) [18, 19]. The studies have indicated projected decline in mean annual rainfall by −3.1, −12.1 and −20.2% in year 2020, 2050, and 2080, respectively.

Cocoa is sensitive to changes in climate conditions for example, sunshine hour, temperature, rainfall, soil conditions and their effects on evapotranspiration. Drought affect growth and yield of cocoa, and the pattern of cropping of cocoa is related to rainfall distribution. Under the projected climate change from 2020 to 2080, cocoa yields may decrease significant below its current yield levels.

It is reported that climate change had produced shifts in the geographical distribution of host and pathogen/pests, altered crop yields and crop loses, stages and rates of development of cocoa pests and pathogens while modifications in host resistance and physiology of host-pathogen/pests interaction may change. Extreme climate induces drought and tree mortality events worldwide, and further increases in tree mortality events are predicted under future climate. Tree mortality enhanced by drought is also noticeable even in moist and wet tropical forests [20, 21]. Climate change is expected to exacerbate mortality events in tropical forest species especially the fruit tree species. Within ecosystems, plant species respond to drought and high temperature stresses of the wet-dry season transitions, environmental (climate) changes will affect ecosystem and plant distribution.

Climate projections using Global Climate Models and change scenarios (SSP-RCP2.5, 4.5 and 8.5), have identified important shifts and shrinkage of suitable areas for tropical crops [19, 22]. The change projections have indicated significant areas of West Africa are likely to experience unfavorable climatic conditions by 2050 [19]. Other reports have affirmed the shifting agroecological landscapes for cocoa in Ghana, Cote D'Ivoire [22]. These reports indicate the need for adaptation planning to ensure sustainability of crop production under current and future climate scenarios [23].

The projections of climate scenarios are relevant for assessing vulnerabilities and impacts of climate change on agriculture, also useful for identifying adaptation strategies capable of sustaining crop production under future climates. Such projections have been made using CORDEX RCP 4.5 and 8.5 in some cocoa producing regions of West Africa [24]. Results showed that critical temperatures thresholds (33°C) would be exceeded which will result in decline in cocoa performance.

There has been declines in quantity and quality of land and accelerated nutrient depletion within the cacao growing agroecology of West Africa particularly Nigeria.

The ecological transect of wet/moist to dry forests of the humid tropics is characterized by increasing frequencies and intensities of droughts and dry spells, water resource scarcity, high temperatures, soil and atmospheric aridity, land degradation (soil erosion and fertility depletion), insect pests and disease pressures. The changing climate and agroecological conditions (vegetation and soil) constitute constraints to sustainable cocoa production in the West African rainforest belt.

5. Climate and soil requirements for cocoa

Cocoa is essentially a plant of the tropical rainforest, it grows well within about 20° north and south of the equator. The equatorial environment with high temperatures and heavy rainfall (between 1500 and 3500 mm) and nitrogen-rich soil is best agroecology for cocoa.

Recommended weather conditions for cocoa especially temperatures: maximum temperatures of 30–32°C and 18–21°C and absolute minimum of 10°C. Light use efficiency relates to temperature because temperatures below 24°C is reported to decreasing radiation use efficiencies especially for photosynthesis rate at light saturation [25]. Cocoa plant has low light saturation point (LSP) of 400 μ E m^{-2} s^{-1} and low maximum photosynthetic rate (7 mg dm^{-1} h^{-1}) at light saturation. It is reported that crop photosynthesis rate decreases if the photosynthetic apparatus is exposed to light intensities exceeding 60% of full sunlight (1800 μ mol m^{-2} s^{-1}), and prolonged exposure to high light intensities damages the photosynthetic apparatus of leaves [25]. Low light intensities however suppress flower and pod production in cocoa. Opeke [8] and Wood and Lass [26] reported that cocoa optimum growth occur under rainfall of 1250–3000 mm per annum (preferably between 1500 and 2000 mm), dry season not longer than 3 months, maximum temperature between 30 and 32°C, minimum of 18–21°C, and no strong winds. Reports from studies under controlled-environment facility (growth chamber) have showed that cocoa performed well under high humidity (80–95% RH) and about 27°C [25].

Cocoa does well on deep soil of loamy sand texture, friable clays, red or reddish-brown in color and soil pH greater than 6.0 [26, 27]. Cocoa can be successfully grown on heavy clay soil, yellow to red overlying a deposit of hydrated iron oxides. Reports from other studies from analyses of soils from the different countries tended to fall into *Alfisols* and *Ultisols* classification [26, 27]. Commonly used terminologies such as "ideal cocoa soils," "ideal cocoa climate" and "marginal cocoa climate" to describe the suitability of agroecologies for its production.

In Nigeria, cocoa cultivation is restricted to south western and eastern parts of the country where the annual rainfall is above 1200 mm. Thus, cocoa production activities are concentrated along rainforest zone of southern Nigeria across in Ondo, Cross River, Ekiti, Osun, Oyo, and Ogun States where the environmental conditions are most suitable. The major cocoa producing areas lie between Latitude 5° 32′ to 7° 47′N and between Longitude 3° 55′ to 8° 42′E (**Figure 1**). The most important varieties of cocoa are *Theobroma cacao* (officially named in 1753 by the Swedish scientist Carl von Linné) and Criollo, Forastero and Trinitario (with numerous hybrids for each strain).

Soils of the south western Nigeria have developed form metamorphic rocks of the basement complex, majority of these soils are of Pre-Cambrian age [27]. The soils are mainly classified as Typic Kanhaplustalf and Typic Haplustalf. The soils of the south eastern Nigeria are derived from basalt under humid tropical forest vegetation and are predominantly classified as Typic Tropohumult (**Figure 2**) [28].

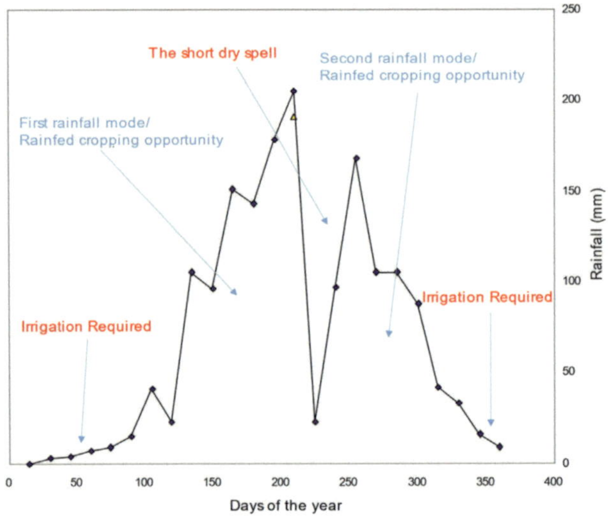

Figure 1.
Bimodal rainfall peaks of the rainforest zone of Nigeria: Rainfed and irrigation based cropping opportunities.

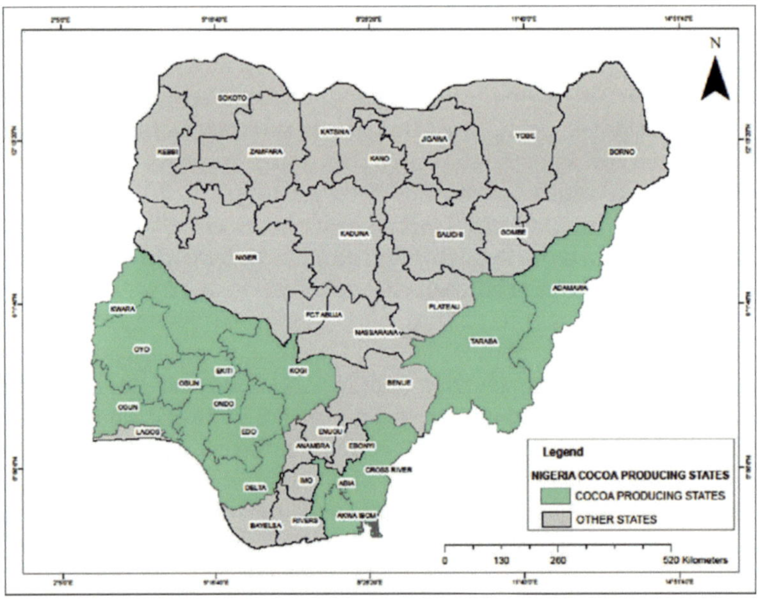

Figure 2.
Cocoa producing states in Nigeria.

6. Climate and agro-ecological conditions

Climate factors play a primary role in determining the ecology of a region [29, 30]. Ecological conditions have profound impacts on the types and scales of agricultural activities [30, 31]. The vegetation zones of Nigeria, are determined by the prevailing climatic conditions, the zones differ in amount and distribution of rainfall, humidity, temperature, atmospheric pressure. The zones are different in annual rainfall and temperature, and predominant species composition (**Figure 3**). These zones have

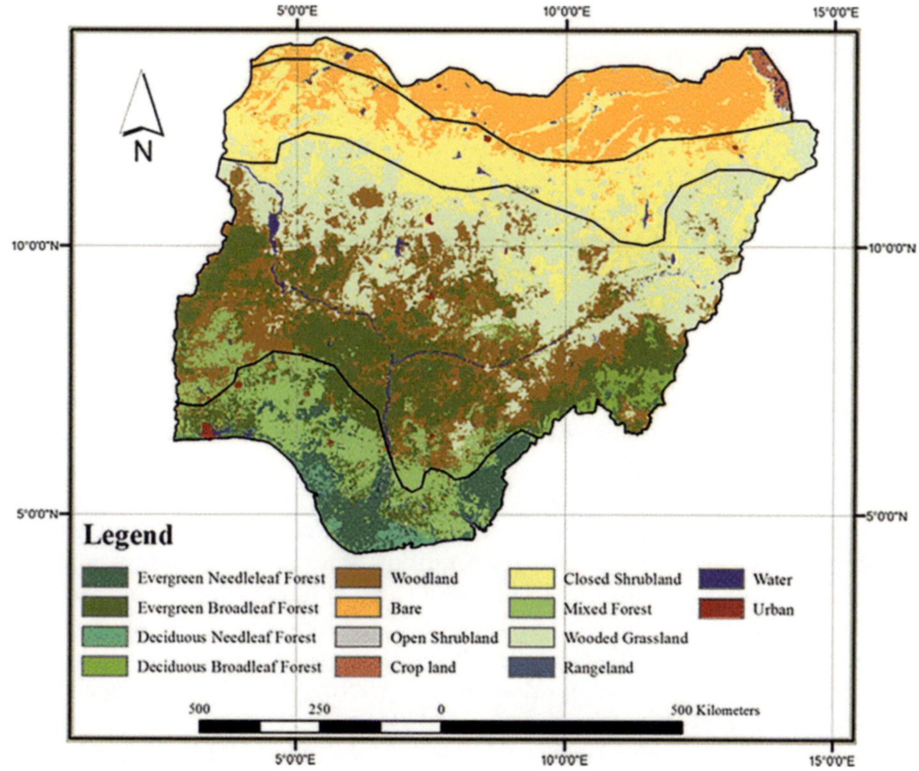

Figure 3.
Vegetation zones of Nigeria.

varying annual rainfall, temperature, atmospheric pressure, and predominant vegetation (dominant species).

Agro-Ecological Zonation (FAO-AEZ) identifies the land's suitability for crop farming and helps understanding of impacts of climate change on agriculture. Agroecological zonation has been used as a tool for identifying agricultural activities under current climate conditions and for predicting future area suitability for crops (FAO (1978, FAO/IIASA (2012) classified Africa into Agro-Ecological Zones (AEZs) based on temperature, precipitation, and soil moisture conditions. Agroceological zonation provides clear picture of land potentials of land resources for use through matching of production system requirements with the land characteristics within the agroecology. The most vulnerable crop producing zones to climate change can be delineated using the concept of AEZs [29, 32]. Modeling results have affirmed shifts in the AEZs under different climate change scenarios [33] and analysis of AEZ had established that climate change affects African landscapes and agroecologies. Results of such analyses for West Africa establish that that the moist (humid) and dry savanna are more vulnerable to climate change while the humid forest or sub-humid AEZs will become more productive in the future [33].

Image analysis using spot-vegetation sensors have showed changes in the composition and extent of different vegetation zones across Nigeria. Reports have confirmed the encroachment of the savanna into the forest belt of southern part of the country. The expansion of non-forest areas, fragmentation and shrinkage of forested ecosystems have ecological and socio-economic consequences [22]. Unsustainable anthropogenic activities such as lumbering, open cast mining, and agriculture have accelerated the degradation of vegetation, soil quality, biodiversity loss, ecosystem services and climate change. The changes in the suitability of agroecologies and crop growing under the present and the future climate using model projections have affirmed differences in suitability ranges from moderate to marginal suitability and the very unsuitable.

7. Climate and agroecological suitability for cocoa

Land suitability classification relates to comparison of requirements of land-use types with properties of land units. Land suitability assessment is a valuable tool for rational soil use planning and sustainable land management.

Agroecological suitability is a measure of the ability of an agroecology to support intended uses (agricultural activities involving production of arable (annual) and permanent (plantation) crops.

Suitability is indicated separately for each land-use type, showing whether the land is suitable or not suitable. Classification of current crop suitability refers to land suitability for a defined use in its present condition, without major improvements. Crop suitability describes appropriateness of an area of land based on the growing threshold of a crop in relation to climatic conditions (minimum and mean monthly temperature and total monthly rainfall) [19, 34]. Crop suitability is also relate to the spatial appropriateness and distribution of land area based on the growing climatic suitability threshold of a crop over time period [35].

Multiple global circulation models (IPSL-CM5A-LR, BCC-CSM 1.1 and MIROC-ESM-CHEM) have been used to simulate changes in crop growing area suitability across the agroecological zones of West Africa future climate change scenarios (RCP 2.6, 4.5, and 8.5) [19, 36]. Areas with increasing and decreasing land suitability were predicted to increase with time, the changes were greater for RCP 4.5 while RCP 8.5 gave the worst prediction indicating higher risk of crop cultivation in the future [36].

Climate departure defines a shift in the climate pattern of a region outside the range of historical variability and such description may be based on local temperature exceeding historical high. Mora et al. [37] described climate departure as the year in which the average temperature of the coldest year after 2005 was warmer than the historic hottest year at a given location. Climate departure manifests as deviation from the historical mean and/or variance of the local climate of an area or region induced by global warming. The authors suggested that West Africa will experience a climate departure based on temperature rise will occur about two decades (2029) earlier than the global mean temperature (2047). The study established the likelihood of changes in large-scale crop and growing area suitability across West African agroecologies.

The concept of crop-climate departure (CCD) is useful for evaluating future changes in crop suitability over historical and future time periods. Crop-climate departure has been used to evaluate future changes in the crop suitability and planting month for crop species in some west African agroecologies [19]. The authors used using Global climate model simulations downscaled by the CORDEX regional climate model (RCA4) to drive the crop suitability model [36]. Results showed a reduction (negative linear correlation) and expansion (positive linear correlation) in area and crop suitability for the guinea and southern sahel zones of West Africa. The study recommended options for short and long-term adaptation and planning for future changes in the crop suitability and planting windows for improving food security and livelihood in West Africa savanna agroecologies.

Crop distribution over agroecologies is predicted using species distribution models (SDMs). Species distribution models are built on genetic algorithm for setting rule for production (GARP) and CLIMEX, and maximum -entropy (MaxEnt) [38–40]. Among the species distribution models, MaxEnt is the most widely used model in recent years [41, 42]. Predictions of impacts of climate (current and future scenarios) on area (agroecology) suitability for crops using Global Circulation Models and adapted MAXENT model. MaxEnt has shown outstanding predictive performance compare with other modeling methods [43, 44]. The simulation of present and potential area suitability or crops in agroecologies and under future climate (2030s, 2050s and 2080s) using MaxEnt model has indicated variabilities in current regional and global agricultural crop area suitability from medium to high suitability in addition to potential undeveloped suitable areas [36]. The predictions also affirmed shifts in the current distribution of crop growing area especially under future climates (2050 and beyond). Especially for West Africa, the studies showed that some current agroecologies of West Africa will be suitable for crops while some others will become unsuitable.

In the rainforest of West Africa, (the cocoa growing belt of the region), predictions have affirmed that yearly and monthly rainfall will decrease slightly by 2050 (aside the coastal areas) and yearly in addition to increases in monthly minimum and maximum temperatures by 2030 with continue increases up till 2050 [19, 36]. In general, West African climate will become less seasonal due to within the year variations, increases in temperature in specific districts by about 1.2°C by 2030 and 2.1°C by 2050 and less seasonal precipitation with decreases in number of dry months (from 4 to 3 months). This trend may imply changes in the suitability and distribution of growing areas within the current cocoa-growing areas in some parts of West Africa (Ghana, Côte d'Ivoire and Nigeria) by 2050 due to temperature increases.

The changing landscapes (agroecologies) and may require changes in agricultural practices as adaptation to new environmental conditions that will prevail while the climatic suitability for growing crops in some parts of West Africa rainforest belt. The development of site-specific adaptation strategies will reduce the vulnerability of smallholder farmers to climate change challenges. The imports of progressive changes in climate and the suitability of crop growing area in particular, with the current areas becoming unsuitable for certain crops may require farmers to shift to alternative crops.

The suitability or otherwise of cocoa growing areas in the rainforest of Nigeria under present and the two future climate models (HadGEM2 and CNRM-CM5 under RCP 4.5 and 8.5) is presented in **Figure 4**. The results showed that compare with present day climate and growing area for cocoa, the unsuitable areas will increase under both scenario predictions and higher magnitude of change for CNRM-CM5 RCP 4.5 and 8.5. greater decline in area suitability is projected for RCP 4.5 for both HadGEM2 and CNRM-CM5.

8. Climate change and distribution range for species

Climate change acts as a major cause of species extinction by impacting the distribution and abundance of species. Species may either keep their current range or respond to changing environmental conditions with range expansions, contractions or shifts [45, 46], which may ultimately contribute towards shrinkage in the forest cover and crucial biodiversity loss [47]. The non-significant changes in habitat suitability for species under climate change scenarios may indicate higher climate resilience. Climate warming can provoke species attrition (changes in species richness) in various ecologies, this may cause increases in colonization of new suitable

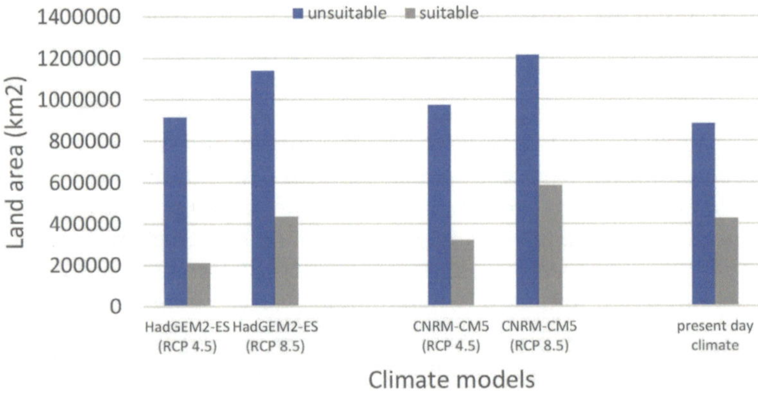

Figure 4.
Fruit tree species (including cacao) land areas of the rainforest zone of Nigeria under the present and future climate models.

areas (expansion range for species) or retractions from unsuitable sites with harsher environmental conditions. Such phenomenon may lead to local and even global species extinction events [46, 48].

9. Environmental factors driving crop growing area suitability

Environmental factors (weather and soil) drive changes in area suitability for crops in agroecologies (AZEs). Climate suitability relates to levels of climate variables which determine growing areas having the potential for successful cultivation and growth of certain crop (s) [45, 46]. Such bioclimatic variables include monthly temperature and rainfall, seasonality (annual range in temperature and precipitation) and extreme or limiting weather factors such as temperature of the warmest month, and rainfall of the wettest and driest months useful for generate biologically meaningful indicators for ecological niche modeling.

Temperature increases had been reported as a major driver of shifts in area suitability for crops under current and future climate. Reports identified temperature of the warmest month as the main driving factor. Temperature explains about 30% of the negative change in area suitability (under 2.4°C increases in temperature of the warmest month by 2050). Temperature seasonality has been identified as the main driving factor for negative change in suitability. The contribution of different bioclimatic variables to changes in crop growing area suitability between current and future climate scenarios, had shown decreasing and increasing trends for crop suitability in agroecologies.

10. Changes in global climate

Global warming denotes the unusually rapid increase *in* Earth's average surface temperature over the *past* century primarily due to the greenhouse [49]. Over both the last 140 years and 100 years, estimates have shown that global average surface temperature has increased by 0.6 ± 0.2°C that is by approximately 0.5–1.0°F (0.3–0.6°C) over the last century [50]. Trends in global average surface temperature between 1993 and 2022 in degrees Fahrenheit per decade confirmed that most of the planet is warming (**Figure 4**). Only a few locations, mostly in

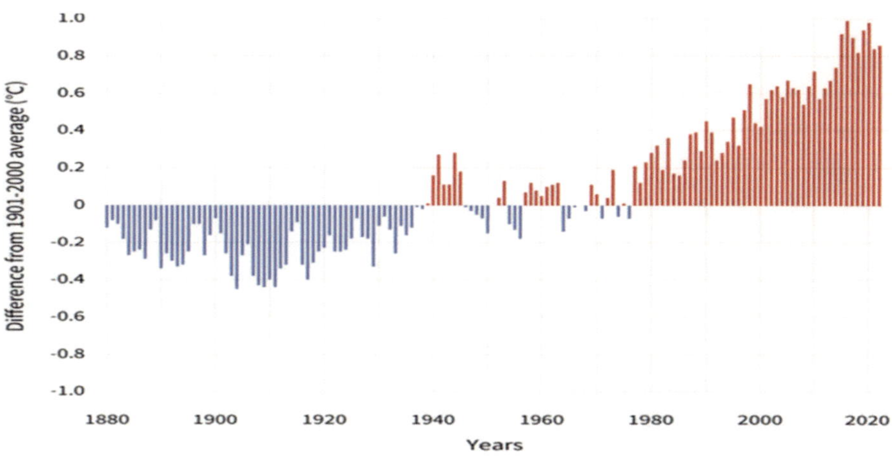

Figure 5.
Global average surface temperature: https://www.climate.gov/media/15022. The average yearly surface temperatures from 1880 to 2022. The cooler-than-average years are depicted in blue bars and warmer-than-average years are shown in red bars. (Source: NOAA Climate.gov graph, based on data from the National Centers for environmental information).

Southern Hemisphere and oceans have cooled over this time period [51]. Trends of increased warming since 1981, has been twice as fast at 0.32°F (0.18°C) per decade [52]. The earth's temperature has risen by an average of 0.14° Fahrenheit (0.08°C) per decade since 1880, or about 2°F in total. The 10 warmest years in the historical record have all occurred since 2010. Over the twentieth-century average of 13.9 and 1.06°C warmer than the pre-industrial period (1880–1900), the year 2022 had warmer surface temperature (0.86°C/1.55°F) (**Figure 4**). Each month of 2022 ranked among the ten warmest for that month (Global Climate Report from NOAA National Centres for Environmental Information, [51]). The "coolest" month was November, which was 1.35°F (0.75°C) warmer than global average (**Figure 5**).

11. The changing climate: the case of Africa

Based on the reports of [50] and Zougmoré et al. [53], over the coming decades, warming from climate change is expected across almost all the Earth's surface while global mean rainfall will increase. In particular, climate change constitutes increasingly serious threat for Africa, a region that has been described as among the most vulnerable continents to the challenges of climate change. Africa is warming faster than the rest of the world on average, records showed that surface temperatures have generally increased over Africa since the late nineteenth century to the early twenty first century by about 1°C, (**Figures 6** and 7). Based on climate projections, many African countries and regions, this will severely compromise agricultural production and food security [50, 54]. Omotosho et al. [55] reported that in West Africa, seasonal cycle of rainfall is driven by the south-north movement of the Inter-Tropical Convergence Zone (ITCZ). The ITCZ is characterized by the confluence between moist south-westerly monsoon winds and the dry north-easterly.

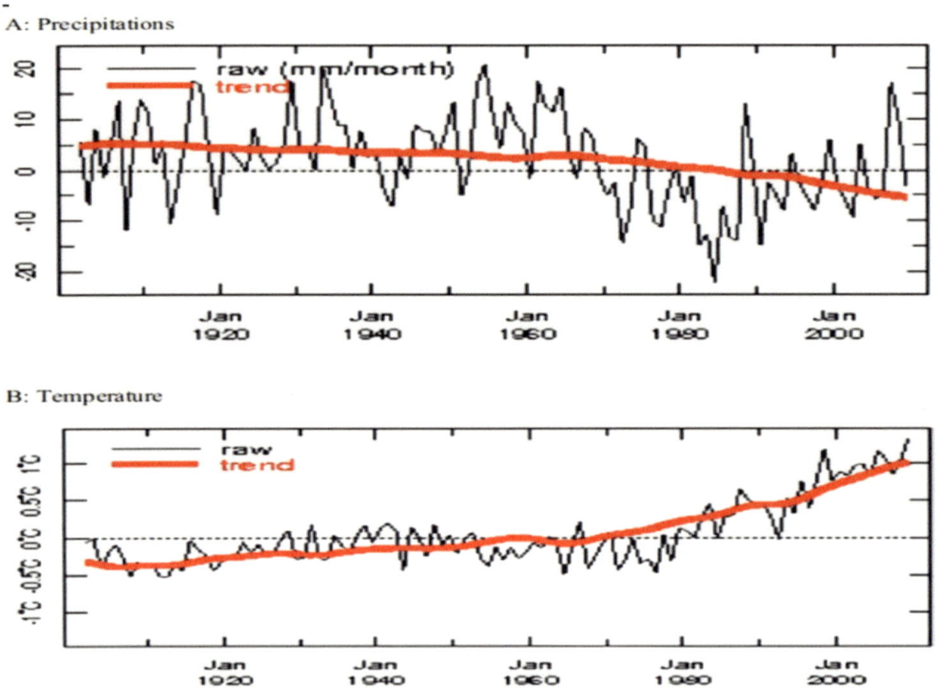

Figure 6.
Variations of the Earth's surface temperature over the last 140 years and the last millennium. Based on records of changes in precipitations (A) and temperature (B) in Africa from 1920 to 2000. (The international research Institute for Climate and Society, Columbia University, Ne. (after [53]).

12. Rainfall and temperature anomalies of the rainforest zone of Nigeria (1980–2020)

The time series (year: 1980–2020) of a normalized annual departure of rainfall and temperatures standardized mean rainfall anomaly, δ) for the rainforest agro-ecological zone of Nigeria. moderate fluctuation in temperature and rainfall are represented by anomaly values of +5 while -5δ denote drought years, and above +0.5 as likelihood of flood events (**Figure 7a**). The normalized annual departure of

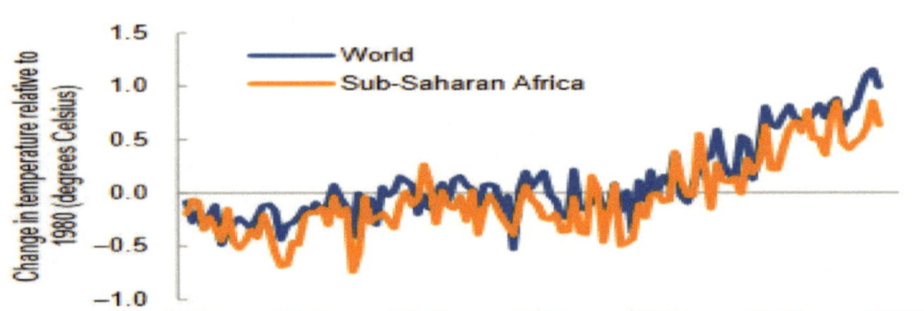

Figure 7.
Changes in global and sub-Saharan Africa (1900 to 2020). a. Rainfall anomaly trends (1980–2020) for rainforest zone of Nigeria. b. Maximum temperature anomaly trends (1980–2020). c. Minimum temperature anomaly trends (1980–2020).

temperatures (maximum and minimum temperatures) is illustrated in **Figures** 7 and **7b**. The results confirmed elevated temperatures characterized by high maxima and minima values which mean high probability of exceeding crop-specific high temperature thresholds. Temperature warming may elicit adjustment of plant physiological functions and water use. Such plasticity is important to acclimation to environmental stresses especially in plant species useful in the revegetation of degraded urban lands. The observed trends confirm changes in climate over Nigeria (for rainfall, maximum and minimum temperatures) for years 1980–2020).

13. Environmental factors driving growing area suitability

Bioclimatic variables are derivable from WorldClim database, from monthly temperature and rainfall values. These Bioclimatic variables are biologically relevant and are often used in ecological niche modeling. Environmental factors (weather and soil) driving change in area suitability for cocoa in the West Africa rainforest belt are exemplified by changes in mean annual temperature and rainfall), seasonality (annual range in temperature and precipitation) and extreme or limiting weather such as temperature of the warmest month, and rainfall of the wettest and driest months. The contribution of bioclimatic variables to the predicted shift in growing area suitability for cocoa between current and 2050 climate scenarios had showed decreasing suitability of weather, soil and vegetation (agroecologies) for cocoa in the tropical rainforest belts of West Africa. Temperature increases had been reported as a major driver of shifts in area suitability for crops especially the changes in suitability between current and future climate especially, the temperature of the warmest month as the main driving factor (it explains about 30% for negative change in area suitability [24]. Climate change will also increase the pressure on forest areas and on other important habitats for fauna and flora [24].

Projections for future climate change suggest a warmer and drier climate in the West African rainforest belt. Information and knowledge improvements with respect to effects of these changes on cocoa production and area suitability in the West African rainforest biome. Simulations using models for the suitability of cocoa's geographical distribution using ensemble of correlative models and projections of two future climate scenarios (RCPs 4.5 and 8.5) by 2050 had been conducted. The models generated information on climate and soil suitability for cocoa. The current and future suitability model had indicated how cocoa production may respond to climate change, and had suggested that reduction in precipitation and increases in temperature may result in reduction in the suitability of the West African rainforest belt cocoa production. The areas suitable for cocoa plantation will decrease by about 37.05 and 73.15% (area suitability for intensification and expansion) under RCP 4.5 and 8.5, respectively, compared with the current climate. Model results also suggest that reduction in precipitation and increase in temperature which may produce a reduction in agroecological suitability for cocoa production in the West African rainforest.

14. Conclusions

Global environment change including climate change has produced marginal environments (agroecologies, weather, soil) and thus major shifts in the suitability of agroecologies for crop production. Thus, climate change has created new environmental boundaries occasioned by changes in rainfall patterns and temperature regimes, seasonal shifts and habitat range for crops. Such changes are expected to affect the area suitable for agriculture, crop species are bound to grow in areas where they were not previously grown and areas that were hitherto not suitable for their production. Climate change may bring about decreases in area suitability for agriculture, and decreases in length of growing seasons and crop yield potentials.

The changing environment conditions are projected to limit cocoa production while the expansion of its production has placed increasing pressure on forest resources in particular along the ecological transect of wet to dry forests. Increase in mean annual temperature up to 2°C will cause considerable decrease in suitability of the rainforest belt of West Africa for cocoa production.

Results of agroecological zoning quantification, show a promise for the extension of suitable areas for the intensification and expansion of cocoa cultivation under future climate. However there will remain, some tracts of land with high levels of soil and climate suitability in the rainforest for cocoa production. Although, tropical crop species are naturally adapted to warmer climates and are developing increasing resilience to climate-related stresses.

Increases in temperatures and declines in rainfall will lead to large parts of sub-Saharan Africa becoming unsuitable for crops. This situation may necessitate a transition to more heat and drought resistant crops to ensure food and nutrition security in the sub-region. The new climatic regimes set by global climate change will affect water resources, agriculture, food and nutrition security, livelihoods and economic growth. In future climates, 2050 and beyond, high percentage loss in the suitability of the West African rainforest for crop production is envisaged. There is however, increasing need to extend production and thus, frontier of fruit tree cultivation to meet rising global demand along other benefits (food security, livelihood, ecosystem services). Sustainable management practices would be required for enhanced productivity and climate mitigation in the era of shifting cacao agroecology and landscapes.

Author details

Samuel Ohikhena Agele[1*], Olufemi Samuel Ibiremo[2] and Oladitan Titilayo[3]

1 Plant Physiology and Ecology Group, Department of Crop, Soil and Pest Management, Federal University of Technology, Akure, Nigeria

2 Cocoa Research Institute of Nigeria (CRIN), Ibadan, Nigeria

3 Department of Crop, Soil and Pest Management, Olusegun Agagu University of Science and Technology, Okiti Pupa, Nigeria

*Address all correspondence to: ohiagele@yahoo.com

References

[1] International Cocoa Organization (ICCO). Cocoa Market Reports. 2022. Available from: www.icco.org

[2] Agele S, Bolarinwa A. Trends of changes in weather and crop yields in Nigeria between 1960 and 2013: Implications for food. Security Research & Reviews: Journal of Ecology and Environmental Sciences. 2018;**10**:21-33. p-ISSN:2347-7822

[3] Agele S. Global warming and drought, agriculture, water resources, and food security: Impacts and responses from the tropics. In: Leal FW, Luetz J, Ayal D, editors. Handbook of Climate Change Management. Cham: Springer; 2021. DOI: 10.1007/978-3-030-22759-3_183-1

[4] Agele S, Charles OA, Agbona A. Oil palm-based cropping systems of the humid tropics: Addressing production sustainability, resource efficiency, food security and livelihood challenges. In: Recent Advances, New Perspectives for Oil Palm. London, UK: Intech Open Publishers; 2021. DOI: 10.5772/intechopen.98257

[5] Laderach P, Valle AM, Schroth G, Castro N. Predicting the future climatic suitability for cocoa farming of the world's leading producer countries, Ghana and Côte d'Ivoire. Climatic Change. 2013;**119**(3-4):841-854. DOI: 10.1007/s10584-013-0774-8

[6] Schroth G. Läderach P. Martinez-Valle IA. Bunn C. Jassogne L. Vulnerability to climate change of cocoa in West Africa: Patterns, opportunities and limits to adaptation. Science of The Total Environment, 2016;**556**:231-241. DOI: 10.1016/j.scitotenv.2016.03.024

[7] Challinor AJ, Watson J, Lobell DB, Howden SM, Smith DR, Chhetri N. A meta-analysis of crop yield under climate change and adaptation. Nature Climate Change. 2014;**4**:287-291

[8] Opeke LK. Tropical Commodity Crops. Ibadan, Nigeria: Spectrum Books Ltd.; 2006. 213 p

[9] Famuwagun B. Agele S, Aiyelari P. Shade effects on growth and development of cacao following 2 years of continuous dry season irrigation. International Journal of Fruit Science. 2017;**18**(7):1-24

[10] Agele S, Famuwagun B, Ogunleye A. Effects of shade on microclimate, canopy characteristics and light integrals in dry season field-grown cocoa (*Theobroma cacao* L.) seedlings. Journal of Horticulture. 2016;**11**(1):47-56

[11] Agele S, Adejobi K, Charles F, Ogunleye A, Olayemi L. Impacts and feedbacks of land use and land cover patterns in landscapes on ecosystem processes and microclimate: Case of a cacao-based agroforestry landscape. Current Journal of Applied Science and Technology. 2017;**22**(3):1-11

[12] Charles EF, Agele SO, Aiyelari OP, Famuwagun IB, Faboade E. Shade and irrigation effects on growth, flowering, pod yields and cacao tree survival following 5 years of continuous dry season irrigation. International Journal of Environment and Climate Change. 2020;**10**(7):54-64. DOI: 10.9734/ijecc/2020/v10i730211

[13] Swiss Platform for Sustainable Cocoa. Produced in the South, Consumed in the North. Bern, Switzerland; Available from: https://www.kakaoplattform.ch/about-cocoa/cocoa-facts-and-figures. [Accessed: May 13, 2023]

[14] Cacao Net. A global strategy for the conservation and use of cacao genetic resources, as the foundation for a sustainable cocoa economy (Laliberté, B.Compil.). Biodiversity International. Sep 2012. ISBN: 978-92-9043-923-3

[15] CBI. The European Market Potential for Certified Cocoa, CBI. 2020. Available from: https://www.cbi.eu/market-information/cocoa-cocoa-products/certified-cocoa/market-potential

[16] Afolayan OS. Cocoa production pattern in Nigeria: The missing link in regional agro-economic development. Analele Universității din Oradea, Seria Geografie. 2020, 2020;**30**(1):88-96. DOI: 10.30892/auog.301110-815

[17] De Alvim PT. Cacao. In: Kozlowski TT, editor. Ecophysiology of Tropical Crops. Elsevier; 1977. pp. 279-313

[18] Akinseye FM, Adam M, Agele SO, Hoffmann MP, Traore PCS, Whitbread AM. Assessing crop model improvements through comparison of sorghum (*Sorghum bicolor* L. Moench) simulation models: A case study of West African varieties. Field Crops Research. 2016;**201**:19-31

[19] Egbebiyi TS et al. Assessing future spatio-temporal changes in crop suitability and planting season over West Africa: Using the concept of crop-climate departure. Climate. 2019;7(9):102. DOI: 10.3390/cli7090102

[20] Agele SO, Olayemi LU, Aiyelari OP, Adesida A. Residual effects of moisture stressed seedlings in the nursery on field performance of cacao genotypes. International Journal of Forestry and Horticulture. 2019;**5**(3):19-30. DOI: 10.20431/2454-9487.0503003

[21] McDowell NG, Pockman WT, Allen DD, Breshears N, Cobb T, Kolb T, et al. Mechanisms of plant survival and mortality during drought: Why do some plants survive while others succumb to drought? New Phytologist. 2008;**178**:719-739

[22] Abu I-O, Szantoi Z, Brink A, Robuchon M. Thiel M. Detecting cocoa plantations in Côte d'Ivoire and Ghana and their implications on protected areas. Ecological Indicators. 2021;**129**:107863. DOI: 10.1016/j.ecolind.2021.107863

[23] Abdulai AR, Ryan Gibson B, Evan D, Fraser D. Beyond transformations: Zooming in on agricultural digitalization and the changing social practices of rural farming in Northern Ghana, West Africa. Journal of Rural Studies. 2023;**100**:103019

[24] Chemura A, Schauberger B, Gornott C. Impacts of climate change on agro-climatic suitability of major food crops in Ghana. PLoS One. 2020;**15**(6):e0229881. DOI: 10.1371/journal.pone.0229881

[25] Daymond AJ, Hadley P. The effects of temperature and light integral on early vegetative growth and chlorophyll fluorescence of four contrasting genotypes of cacao (*Theobroma cacao*). Annals of Applied Biology. 2004;**145**(3):257-262

[26] Wood GA, Lass RA. Cocoa, Tropical Agricultural Series. 4th ed. London: Longman; 1985: 620 p

[27] Smyth A, Montgomery RF. Soil and Land Use in Central Western Nigeria. Government Printer: Ibadan; 1962. p. 265

[28] Soil Survey Staff. Soil taxonomy. A basic system of soil classification for making and interpreting soil surveys. In: Agricultural Handbook 436, Natural Resources Conservation Service. 2nd ed. Washington DC, USA: USDA; 1999; 869 p

[29] Seo SN. Evaluation of the agro-ecological zone methods for the study of climate change with micro farming decisions in sub-Saharan Africa. European Journal of Agronomy. 2014;**52**(B):157-165

[30] Zhang P, Zhang J, Chen M. Economic impacts of climate change on agriculture: The importance of additional climatic variables other than temperature and precipitation. Journal of Environmental Economics and Management. 2017;**83**:8-31

[31] Nordhaus WD. A review of the "stern review on the economics of climate change". Journal of Economic Literature. 2006;**45**(3):686-702

[32] Schlesinger WH. Biogeochemistry. In: An Analysis of Global Change. 2nd ed. San Diego, London: Academic Press; 1997; 588 p

[33] Kala N, Kurukulasuriya P, Mendelsohn R. The impact of climate change on agro-ecological zones. Environment and Development Economics. 2012;**17**(6):663-687

[34] Singh D, McDermid SP, Cook BI, Puma MJ, Nazarenko L, Kelley M. Distinct influences of land-cover and land-management on seasonal climate. Journal Geophysical Research Atmosphere. 2018;**123**(21):12017-12039. DOI: 10.1029/2018JD028874

[35] Izumi T, Ramankutty N. How do weather and climate influence cropping area and intensity? Global Food Security. 2015;**4**:46-50

[36] Liu F et al. Modeling the current land suitability and future dynamics of global soybean cultivation under climate change scenarios. Field Crops Research. 2021;**263**(1):108069

[37] Mora C, Frazier A, Longman R, et al. The projected timing of climate departure from recent variability. Nature. 2013;**502**:183-187. DOI: 10.1038/nature12540

[38] Phillips SJ, Dudik M. Modeling of species distributions with MaxEnt: New extensions and a comprehensive evaluation. Ecography. 2008;**31**:161-175. DOI: 10.1111/j.0906-7590.2008.5203.x

[39] Ramirez-Cabral NYZ, Kumar L, Shabani F. Global alterations in areas of suitability for maize production from climate change and using a mechanistic species distribution model (CLIMEX). Scientific Reports. 2017;**7**:5910

[40] Wegier A, Alavez V, Vega M, Azurdia C. *Gossypium hirsutum*. In: IUCN Red List of Threatened Species. Cambridge, United Kingdom; 2019. DOI: 10.2305/iucn.uk.2019-2.rlts.t71774532a71774543.en

[41] Liu C, White M, Newell G. Selecting thresholds for the prediction of species occurrence with presence-only data. Journal of Biogeography. 2013;**40**:778-789. DOI: 10.1111/jpi.12058

[42] Shabani F, Ahmadi M, ,Peters KJ, Haberle S, Champreux A, Saltré F, Bradshaw CJ. Climate-driven shifts in the distribution of koala-browse species from the last inter-glacial to the near future. Ecography, 2019;42:1587-1599

[43] Liang W, Papes M, Tran M, Grant J, et al. The effect of pseudo-absence selection method on transferability of species distribution models in the context of non-adaptive niche shift. Ecological Modeling. 2018;**388**:1-9

[44] Padalia H, Strivastava V, Kushwaha SPS. Modeling potential invasion range of alien invasive species, *Hyptis suaveolens* (L.) Poit in India: Comparison of MaxEnt and GARP. Ecological Informatics. 2014;**22**:36-43

[45] Pramanik M, Paudel U, Mondal B, Chakraborti S, Deb P. Predicting climate change impacts on the distribution of the threatened *Garcinia indica* in the Western Ghats, India. Climate Risk Management. 2018;**19**:94-105

[46] Park I-k, Borzée A, Park J, Min SH, Zhang Y-P, Li S-R, et al. Past, present, and future predictions on the suitable habitat of the slender racer (*Orientocoluber spinalis*) using species distribution models. Ecology and Evolution. 2022;**12**:e9169. DOI: 10.1002/ece3.9169

[47] Attore F, Alfo M, Sanctis MD, Bruno F, et al. Evaluating the effects of climate change on tree species abundance and distribution in the Italian peninsula. Applied Vegetation Science. 2023;**4**(2):242-255

[48] Thuiller W, Lavorel S, JOMB A, Pretice IC. Climate change threats to plant diversity in Europe. Proceedings of the National Academy of Sciences. 2005;**102**(23):8245-8250. DOI: 10.1073/pnas.0409902102

[49] NASA Earth Observatory. 2010. Available from: https://earthobservatory.nasa.gov › features /Global warming

[50] International Panel on Climate Change (IPCC). The Physical Science Basis. Working Group I, Fifth Assessment Report of the Intergovernmental Panel on Climate Change (IPCC). Geneva, Switzerland; 2013. Available from: https://www.ipcc.ch/reports

[51] National Oceanic and Atmospheric administration (NOAA). Global Climate Report. Silver Spring, Maryland, USA: National Centers for Environmental Information. 2022

[52] Climate gov. 2023. Available from: https://www.climate.gov/media/15022.

[53] Zougmoré et al. Modifications des précipitations et des températures en Afrique enregistrées entre 1920 et 2000. York, USA: Columbia University; 2016

[54] International Panel on Climate Change (IPCC). AR4 Climate Change 2007: Synthesis Report Synthesis Report Working Groups of the Intergovernmental Panel on Climate Change (IPCC) during the AR4 Cycle. Geneva, Switzerland; 2007. Available from: https://www.ipcc.ch/reports

[55] Omotosho J, Agele SO, Balogun IA, Adefisan EA. Climate variability, crop-climate modeling and water ecophysiology research: Implications for plant's capacities for stress acclimation, yield production and food security. Global Journal of Plant Ecophysiology. 2006;**3**(2):56-69, 2013

Chapter 2

Potential Impact of Stratospheric Aerosol Geoengineering on Cocoa Suitability in Nigeria

Temitope S. Egbebiyi, Christopher Lennard, Pinto Izidine, Romaric C. Odoulami, Piotr Wiolski and Akintunde I. Makinde

Abstract

Cocoa is an important cash crop that contributes to the economy of Nigeria *via* job creation and foreign exchange earnings. However, escalating global warming trends threatens Cocoa cultivation and have resulted in a decline and heightened variability in Cocoa production in Nigeria, with potential for further exacerbation in the future. A potential way to reduce the warming is through climate intervention (CI) techniques, including Stratospheric Aerosol Injection (SAI), which involves the injection of sulphur into the stratosphere to reflect a small percentage of incoming solar radiation and lower earth's temperature. To gauge GHG and SAI impact on Cocoa suitability in Nigeria, we used Geoengineering Large Ensemble Simulations (GLENS) dataset as input into Ecocrop model for historical (2011–2030) and future periods (2070–2089). Our results show GHG impact will increase mean and minimum temperatures (up to 3°C) and total monthly rainfall (up to 15 mm) by the end of century in the southwest and north-east area of Nigeria while rainfall decrease of similar magnitude in the other parts of the country. With SAI intervention, rainfall may decrease by about 10–20 mm over the country and reduce mean and minimum temperature by 2°C. Suitable land for Cocoa cultivation in Nigeria may decrease by 24 and 18% under GHG and SAI, respectively, while unsuitable may increase by 14 and 24% by the end of century. Our study has implications for the economies based on Cocoa production in Nigeria.

Keywords: cacao, Nigeria, Ecocrop, crop suitability, stratospheric aerosol injection

1. Introduction

Cocoa (*Theobroma cacao*) holds immense economic significance in Nigeria, contributing significantly to the social and economic well-being of the country. With a history spanning centuries, Cocoa has played a pivotal role in shaping Nigeria's economy and fostering rural livelihoods before the discovery of crude oil [1, 2]. Nigeria has achieved significant milestones in Cocoa production, ranked as one of the top four Cocoa-producing nations globally [2, 3], with Cocoa production largely concentrated in the southern regions of the country [1]. The crop has become a cornerstone of Nigeria's agricultural sector, providing a livelihood for millions of smallholder farmers and supporting numerous downstream industries, such as Cocoa processing and chocolate production [4–6]. The foreign exchange earnings from Cocoa exports have also significantly contributed to Nigeria's overall revenue and balanced trade [1]. The Cocoa industry's growth and achievements have made it an essential component of Nigeria's socio-economic fabric. Moreover, its Pan-Africa significance as a tropical African plant cultivated for its oil and bold foliage places indispensable economic and livelihood significance on Cocoa amongst the people engaged in its cultivation [7–10]. However, climate change poses a severe threat to Cocoa cultivation, impacting its suitability and potentially hampering Cocoa-dependent economies.

The potential impact of climate change or global warming on Cocoa production and suitability in Nigeria is a pressing concern that has garnered extensive research attention. A number of studies (e.g., [11]) have underscored the alarming implications of rising temperatures and changing weather patterns on Cocoa yields. Global warming's adverse effects on Nigeria's Cocoa industry are evident, as projected decreases in Cocoa yield pose substantial economic risks for Cocoa farmers [12, 13]. Furthermore, Nwachukwu et al. [14] emphasise the vulnerability of Cocoa productivity to climate change-induced shifts in temperature and rainfall. These changes, combined with limited adaptation measures, could potentially exacerbate the negative effects on Cocoa suitability. Agbongiarhuoyi et al. [7] delve into the perceived effects of climate change, highlighting the concerns of Cocoa farmers about the impact on various aspects of Cocoa cultivation. In light of these findings, adaptation strategies become paramount. Jamal et al. [15] stress the importance of climate adaptation efforts, particularly for small-scale farmers who are disproportionately affected. These studies collectively underscore the significant potential impact of climate change on Cocoa production and suitability in Nigeria, urging the implementation of effective strategies to mitigate its adverse effects.

Solar Radiation Management (SRM) has been identified as a potential technique for climate mitigation efforts [16–18] aimed at reducing CO_2 emissions to limit the increasing global temperature. One of the SRM prominent approaches is the Stratospheric Aerosol Injection (SAI) which involves the release of gaseous aerosol precursors such as sulphur dioxide (SO_2) into the stratosphere to form aerosols that reflect into space a small amount of incoming solar radiation with the aim to reduce temperature at the earth's surface [18, 19]. Although this strategy may decrease global temperature, studies have revealed that it could have a negative impact on rainfall notably in most parts of Africa [20–24]. For example, findings have shown that the impact of SRM will help increase crop yield via the reduction in temperature and heat stress, whilst its resultant effects in rainfall deficit may lead to a reduction in suitable areas for crop cultivation and yield in other regions [16, 20]. Xia et al. [20] also revealed that SRM poses a threat to food security notably in areas like Nigeria, where agricultural productivity is dependent on monsoon rainfall [20, 25]. However, despite this research, there is still a dearth of information on how the SAG will affect Cocoa suitability and yield in Nigeria and the present study is in that direction.

This chapter explores the potential impact of SAI to counteract the adverse effects of climate change on Cocoa cultivation suitability in Nigeria. Section 2 describes the data and methodology on the paper, the results are presented and discussed in Sections 3 and 4, respectively, whilst the summary and conclusions are in Section 5.

2. Data and methods

2.1 Study domain

Our study domain for this research is Nigeria, the most populous African nation and one of the four largest producers of Cocoa in the world (**Figure 1**). It has rainfed agriculture as its mainstay economy and means of livelihood. We define Nigeria domain from latitude 4–14°N and longitude 2.4–15°E with agroecological zones of Nigeria designated as Guinea, savanna and Sahel zones (**Figure 1**). The region is characterized by a strong north-to-south temperature and precipitation gradient [26] and the West African Monsoon Systems (WAMs) are the main rainfall-producing system in Nigeria [27]. The Nigeria climate is in the humid southern area, with rainfall amount up to 3000 mm/year and semi-arid in the north with rainfall amounting to about 450 mm/year [26]. The south experiences a bimodal rainfall regime between March–July and September–November whilst the Northern region experiences unimodal rainfall regime from May–October and agricultural production is highly dependent on rainfall [26]. Different crops are cultivated in different parts of Nigeria and contribute significantly to the economy of the country [26, 28]. Major crops grown in Nigeria include Maize, Yam, Cowpea, Cassava, Rice, Groundnut, Cocoa, Oil palm [29]. Cocoa is reported to be the most important cash crop in West Africa, particularly due to its substantial contribution to the country's GDP via export and foreign earnings [1, 30]. Hence, the need for this present study is to provide information on how climate geoengineering will affect Cocoa cultivation and suitability in Nigeria.

2.2 Data

2.2.1 GLENS datasets

For the study, we analysed the National Centre for Atmospheric Research Community Earth System Model (CESM1) simulation for the Stratospheric Aerosol Geoengineering Large Ensemble Project (GLENS) dataset [31]. The GLENS experiment uses the Whole Atmosphere Community Climate Model (WACCM) as its atmospheric component (WACCM; [32]) with a horizontal resolution of 0.9° latitude × 1.25° longitude and 70 vertical levels from the surface up to 140 km [31]. The experiment includes a multi-member ensemble simulation of future climate and we used two GLENS experiments, the control and the feedback, both experiments were forced with

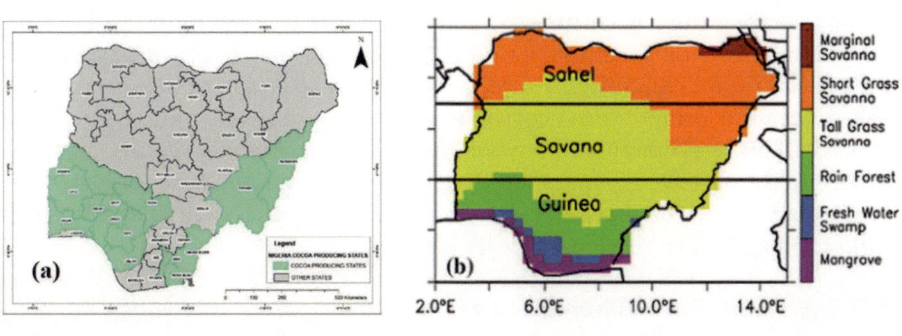

Figure 1.
Map of Nigeria showing (a)agroecological zones designated as Guinea, savanna and Sahel zones and Cocoa producing states in Nigeria (source: [1]).

high-end future greenhouse gas scenario (RCP8.5). However, for the feedback experiment, sulphur dioxide (SO_2) was injected into the stratosphere simultaneously at four different locations, along longitude 180°E and latitudes 15°S, 15°N at 25 km and 30°S and 30°N at 22.8 km over a grid point and between 5 and 7 km above the tropopause to keep near-surface global temperature at the 2020 value under the RCP8.5 emissions scenario [19, 31]. The control experiment has two datasets: the baseline (2010–2030) and the RCP8.5 simulation dataset (2010–2097), whilst the feedback experiment period is from 2010 to 2097. The control experiment baseline datasets and feedback datasets have 20 ensemble members with simulation over the period 2010–2030 and three members extended to the end of century, 2010–2097. Hence, for consistency, we used simulations from the three ensemble members that extend to the end of century.

We analysed 20-year periods in each simulation dataset. Firstly, for the control baseline simulation dataset (hereafter Hist), we used 2011–2030 period to understand the spatial distribution and characteristics of Cocoa suitability distribution during the present-day climate. This period was chosen based on the previous findings by Tilmes et al. [31] and Simpson et al. [24] which defined this timeframe as a target for SAI to keep the surface temperature at 2020 levels under RCP8.5 scenario until the end of century. The control RCP8.5 simulation (hereafter, GHG) dataset was used to evaluate the spatial characteristics of Cocoa suitability under RCP8.5 scenario for the period 2070–2089, whilst the feedback simulation dataset (hereafter, SAI) was used to understand the spatial characteristics of Cocoa suitability over West Africa under RCP8.5 scenario with SAI interventions for the period 2070–2089. We also evaluate the global model capability to represent the current period by comparing the control baseline simulation to CRU-WFDEI suitability output. We quantify the impact of climate change on Cocoa suitability under RCP8.5 scenario by GHG minus HIST. Meanwhile the SAI minus HIST shows the impact of RCP8.5 and SAI on future crop suitability over West Africa and SAI minus GHG shows the SAI intervention of future suitability of Cocoa in the regions. As with previous studies e.g., [22, 23, 33], we used monthly datasets of mean and minimum monthly temperatures and total monthly rainfall.

2.3 Methods

2.3.1 Bias correction

All the GLENS datasets were bias corrected using the Climate Research Unit (CRU) observation-based reference dataset at a horizontal resolution of 0.5° for the period 1980–2009 [34]. A standard quantile-quantile bias-correction approach was employed to correct the two climate variables, temperature (minimum and mean) and rainfall required for our study period 2011–2030 for the historical/baseline period and 2070–2089 for RCP8.5 and SAI simulations [35]. The resultant bias corrected variables were used as input into Ecocrop model for our crop suitability experiments over Nigeria. This step is highly important because climate model simulation often deviates from the observed climatological data. Hence, the need for bias correction before the data is used for climate change impact assessments, such as hydrological modelling and agricultural impact studies [36].

2.3.2 Crop suitability modelling using Ecocrop

The impact of SAI on Cocoa suitability in Nigeria was investigated using the Ecocrop suitability model. Crop suitability was calculated as described in

Ramirez-Villegas et al. [37] based on the crop growth suitability threshold from Food and Agriculture Organisation, FAO-Ecocrop database (**Table 1**) [37, 38]. The model evaluates the relative suitability of crops in response to a range of climates including rainfall, temperature, and the growing season for optimal crop growth and operates at a monthly scale with the capacity to analyse crop suitability across different geographical location [37, 39, 40]. Ecocrop works based on the environmental ranges of a crop coupled with numerical assessment of the environmental condition to determine the potential suitable climatic condition for a crop. The suitability rating can be linked to the agricultural yield which is partly dependent on the strength of the climate signal [37, 41]. We used Ecocrop to produce a monthly suitability index for Cocoa and demonstrate the impact of SAI on its suitability in Nigeria. Crop suitability thresholds are based on a FAO-hosted dataset which acts as a baseline from which to quantify departures in each scenario, under future climate simulation with natural forcing (GHG) and SAI at RCP8.5 using the GLENS ensemble simulation datasets.

2.3.3 Sensitivity to climatic variables, rainfall and temperature

We also independently test the influence of fluctuations and trends in rainfall and temperature on Cocoa suitability. For example, to test the influence of rainfall, we use total monthly rainfall values over the study period for both the historical (2011–2030) and future periods (2070–2089) but keep temperature constant. The constant temperature is from the long-term mean monthly temperature so that the monthly temperature (both minimum and mean temperatures) for January to December each year over the study period with or without SAI is the same. This experiment is called "rain-vary". To test the influence of temperature, a similar approach was used, with the exception that total monthly precipitation values were held constant over the study period, whilst monthly temperatures, both minimum and average, varied from year to year. This experiment is called "temp-vary". When all the three variables (rainfall, mean and minimum temperature) vary simultaneously, the experiment is called "all-vary".

2.3.4 Statistical analysis

The robustness of the projected impact of GHG and SAI on Cocoa suitability and planting season crops over Nigeria was assessed based on the condition that all three simulations agree on the sign of change. Previous studies [23, 39, 40, 42–45] have all used the methods to test and indicate the robustness of climate change signals. We also examined the fractional percentage of suitability for the three experiments over Nigeria by aggregating the different suitability index value at each grid point for the study period.

Crop name	**Growing period (Days)**	**Climate variables**							
		Temperature				**Rainfall**			
Cacao	180–365	Tmin	Topmin	Topmax	Tmax	Rmin	Ropmin	Ropmax	Rmax
		10	21	32	38	900	1200	3000	7600

Table 1.
Cacao growth threshold as generated by the FAO-Ecocrop model.

3. Results and discussion

3.1 Impact of global warming and SAI on mean climate variables in Nigeria

Figure 2 shows the spatial distribution of the total monthly rainfall and temperature variables, minimum and mean temperatures over Nigeria. For example, CESM1 model shows rainfall gradient from south to north in rainfall distribution, as rainfall amount decreases as you move northward over Nigeria. Our simulation shows that the south coast in the Guinea zone of Nigeria receives the highest total monthly rainfall amount, 240 mm/year, whilst the lowest total monthly rainfall amount, 80 mm/year, is to the north. For temperature, our result shows the spatial distribution of both mean and minimum temperatures over Nigeria with about 20°C and 25°C, respectively. The impact of climate change, GHG (RCP8.5) relative to baseline period (2011–2030) varies for both rainfall and temperature in Nigeria (**Figure 2**, column 2). Whilst GHG shows a similar pattern of projected increase in mean and minimum temperature over Nigeria, there is a variation in the projected change in rainfall over the region. For example, an increase of 1°C minimum temperature is expected over the country except in Abuja, the north-central part, where a decrease of temperature of 1°C is projected. In contrast, a projected increase up to 2–3°C warming in mean temperature is expected over the country due to GHG. The projected increase in mean temperature is expected to be about 3°C around the south-east and northern Sahel, whilst a 2°C is expected in other parts of the country. For rainfall, with reference to the historical period, a projected increase in total monthly rainfall up to 25 mm (about 1.25 mm/month) by the end of century relative to the historical over the south-western part of Nigeria and up to 10 mm (0.5 mm/month) and 5 mm (0.25 mm/month) in the north-east and north-west Sahel zone of the country, were observed, respectively. On the other hand, GHG may lead to decrease in 10–15 mm in the south-east and central part of the country. These findings are in line with Pinto et al. [22] and Abiodun et al. [23].

The deployment of SAI shows a reverse pattern in total monthly rainfall and temperature variables in comparison to the baseline period and impact of RCP8.5 GHG over Nigeria (**Figure 2**, column 4). For example, the impact of SAI technique relative to historical period may lead to a decrease of about 1–2°C over Nigeria as it induces a cooling due to reduction in mean and minimum temperature. This implies that the deployment of SAI showed a reduction in climate warming over Nigeria and in total monthly rainfall up to about 5–25 mm relative to baseline period (2011–2030) across the country except the south-west zone of the country with higher magnitude up to 35 mm. Furthermore, we examined the impact of SAI deployment on rainfall and temperature relative to GHG. Our findings show that SAI deployment will lead to a decrease in mean and minimum temperature variables up to 3°C from the impact of GHG thus inducing further cooling over the country [46]. In addition, the projected impact of SAI deployment in comparison to GHG-induced impact shows a reduction in the magnitude of decrease in total monthly rainfall in the south-east zone of the country whilst it induces further drying in the south-west and north-east region of Nigeria.

3.2 Evaluation Cocoa suitability in the historical period

Cocoa suitability varies over Nigeria for the historical period (**Figure 3**, column 1). In general, there is a decreasing suitability gradient from south to north of Cocoa over Nigeria. As a result, Cocoa suitability decreases northward and notably unsuitable in the northern part of Nigeria although with variation in spatial extent.

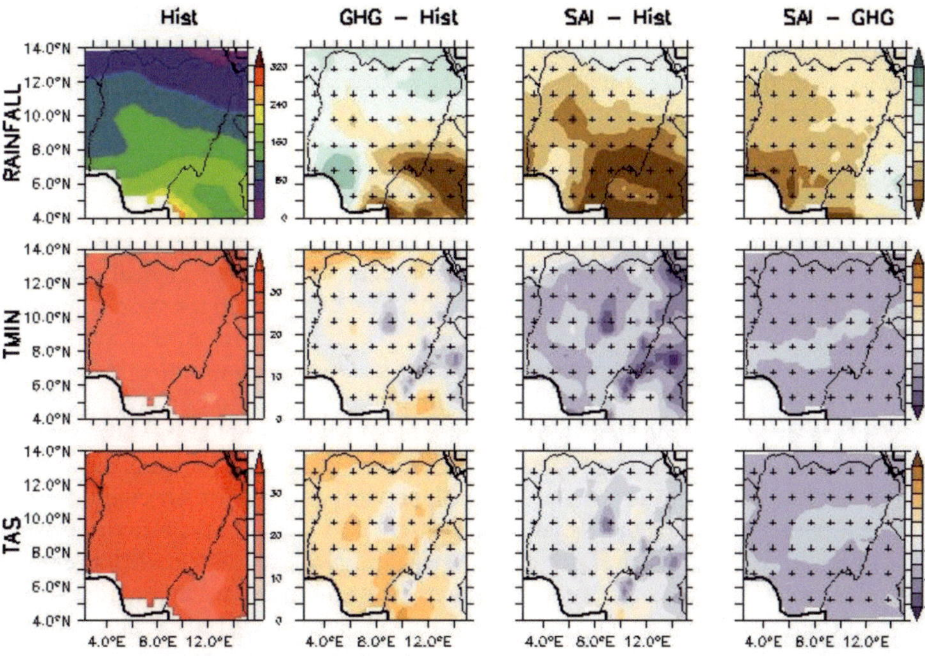

Figure 2.
The spatial distribution of climate variables (total monthly precipitation, mean and minimum monthly temperature) in the present-day climate over southern Africa (first column; 1980–2009) and their projected future changes in the period (2065–2090) under the RCP8.5 scenario without and with SAI (i.e., second and third columns, respectively). The extent to which the SAI influences the impacts of global warming on the variables is presented in the fourth column. The cross sign (+) indicates where at least 75% of the simulations agree on the sign of the changes.

Ecocrop spatial suitability characteristics simulation depicts that a large area to the north of 10°N is unsuitable with Suitability Index Value (SIV) below 0.20 (0–0.20) for Cocoa in Nigeria, notably in Sahel zone in the northern part of Nigeria. However, our result shows that higher suitability (SIV $\geq$ 0.5) of Cocoa is observed to the south of 10 N, notably in the Guinea agroecological zone. In general, the observed data, CRU-WFDEI shows a good agreement with CESM1-WACCM dataset in the spatial suitability distribution of Cocoa over Nigeria (**Figure 3**). The model captures well main Cocoa producing areas in Nigeria, notably the south-west zone (with SIV above

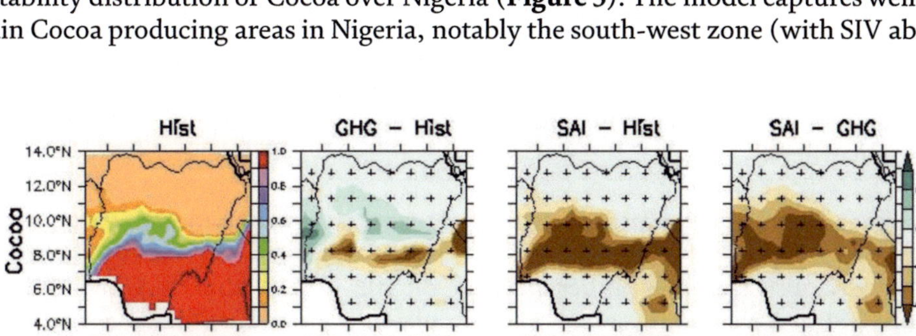

Figure 3.
The spatial distribution of Cocoa suitability for the historical period over Nigeria (first column; 2011–2030) and their projected future changes in the period (2070–2089) under the RCP8.5 scenario without and with SAI (i.e., second and third columns, respectively). The impact of SAI-induced effect on global warming on Cocoa suitability is presented in the fourth column. The cross sign (+) indicates where all ensemble members agree on the sign of projected changes.

0.6) which contributes about 80% of its production. The output shows that there is a strong spatial correlation (r = 1) between CRU-WFDEI and CESM1-WACCM data for Cocoa. Our findings are consistent with the results of Afolayan [1] on the spatial suitability distribution of Cocoa over Nigeria.

3.3 Projected changes in Cocoa suitability in Nigeria

The impact of climate change (GHG) and SAI on Cocoa suitability shows a similar spatial pattern but with varied magnitude across the agroecological zones of Nigeria (**Figure 3**, column 2). GHG as projected may lead to an increase (up to 0.20) in suitability index of Cocoa along the south-west boundary of Oyo and Ogun state and about 0.1 along eastern savanna in Nigeria. On other hand, the projected impact of GHG over Nigeria may lead to a decrease (about 0.2) in Cocoa suitability in the Guinea-savanna zone (lat. 7–9°N). However, no change in Cocoa suitability is expected in the south coast of Nigeria in the Guinea zone, which means the area remains suitable for Cocoa as observed in the historical period. Similar characteristic is also observed in the northern Sahel zone north of 12°N, as the area remains unsuitable as observed in the historical period. These findings are consistent with previous studies e.g., [12, 14, 47] that reduction in rainfall and increases in temperature will affect the suitability and cultivation of Cocoa in Nigeria. Our result also agrees with Schroth et al. [30] that Guinea-savanna zone will be the most affected Cocoa cultivated area in Nigeria under GHG. Thus, reduction in suitable areas may lead to reduction in Cocoa yield and production over Nigeria [14, 48].

The impact of SAI on Cocoa suitability relative to the baseline period (2011–2030) shows a similar spatial suitability pattern as that of GHG but with a variation in the Guinea-savanna zone (**Figure 4**, column 4). With SAI, no change in Cocoa suitability is detected in the Guinea and Sahel zone of Nigeria, suggesting SAI preserves the spatial distribution of current Cocoa cultivation suitability in the future here. On the other hand, a decrease (up to 0.3) in Cocoa suitability is expected over Guinea-savanna zone of Nigeria with SAI, and the magnitude of the projected decrease in Cocoa suitability is expected to be higher with SAI deployment in comparison to GHG-induced impact. In addition, SAI intervention is projected to result in decrease in suitability index value (about 0.35) of Cocoa in most parts of Nigeria. No projected change is expected in southern and north-eastern part of the country, this signifies that the southern zone remains a key region for the cultivation of Cocoa with SAI deployment over the country. However, the north-east in the Sahel zone remains unsuitable for growing Cocoa as observed in the historical period. The general projected decrease in the suitability under SAI may be linked to the projected reduction in rainfall over Nigeria.

3.4 Impact of global warming and SAI on planting season in Nigeria

Our study also examines the best planting months (PM) within the Growing Season (GS) over Nigeria (**Figure 4**). The simulated planting month represents the best month of the planting window and varies across the three Agroecological zones of Nigeria. For the historical climate, Ecocrop simulation shows a variation in the planting windows for Cocoa over Nigeria (**Figure 4**, column 1). Ecocrop shows January as the best PM for Cocoa in Guinea zone south of 7°N and north of 12°N in the Sahel zone, both areas are notable for their suitability in the south and unsuitability in the north, respectively, for the cultivation of Cocoa in Nigeria. Also, March–May

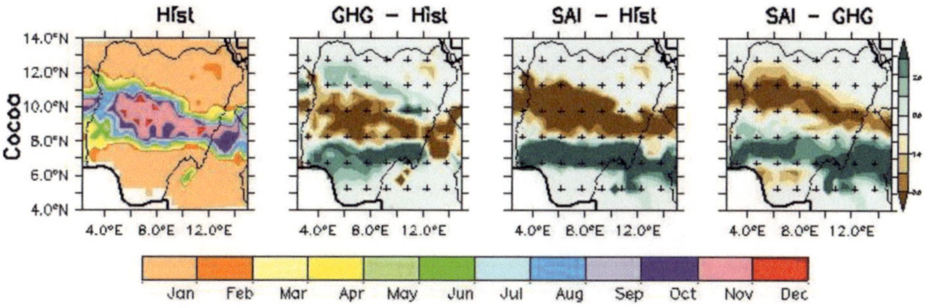

Figure 4.
The spatial distribution of planting season of Cocoa for the historical period over Nigeria (first column; 2011–2030) and their projected future changes in the period (2070–2089) under the RCP8.5 scenario without and with SAI (i.e., second and third columns, respectively). The impact of SAI-induced effect on global warming on Cocoa's planting season is presented in the fourth column. The cross sign (+) indicates where all ensemble members agree on the sign of projected changes.

is the best planting month over the south-west Guinea-savanna zones (lat. 7–10°N), notably in the major producing areas in the southern part of Nigeria. In addition, the months of October–November were observed as the most suitable period in the central extending to south-east savanna zones of the area. The simulated planting months are consistent with the past findings of Afolayan [1] that the months of March–May are the best planting months for Cocoa in Nigeria, notably in the southern part of Nigeria.

The impact of GHG and SAI on Cocoa planting season relative to the historical period shows similar spatial patterns across the different agroecological zones of Nigeria but with variation in magnitude (**Figure 5**, column 2). The impact of GHG may lead to a delayed planting season in the Guinea zone and an area in the Sahel zone of Nigeria although at different magnitude. For example, Cocoa planting season may be delayed by 2 months in south coast of Nigeria in the Guinea zone (south of 7°N) and south-east savanna around lat. 10°N whilst a month delay may be expected from the north-west Sahel zone extending to the central (about 11°N) savanna zone. Hence, GHG warming indicates a shift in Cocoa planting months to March over the Guinea zone and south-east savanna zone and February in the north-west Sahel zone and south-east savanna zone in comparison to the historical period. In contrast, a projected early planting may be expected over the Guinea-savanna zones (7–11°N) for Cocoa in Nigeria except along the south-west boundary between Ogun and Oyo state with one mouth delay. The projected early planting imply that June-July may be the best planting season in the Guinea-savanna zones and March along south-west boundary between Ogun and Oyo state of Nigeria. However, no change in planting season is expected in the north-east Sahel zone of Nigeria, hence, January remains the best planting month over the area. The impact of SAI on the plantings shows similar spatial pattern or characteristics as that of GHG effect relative to the historical period but at a higher magnitude. The projected delay and early planting in the Guinea-savanna and savanna-Sahel zones, respectively, may be up to 3 months under SAI relative to the historical period.

The impacts of SAI intervention (i.e., SAI-GHG) on planting season show a similar spatial characteristic as its impact relative to the historical period but with variation in magnitude (**Figure 5**, column 4). An early planting up to 3 months may be expected in the north savanna zone extending into the Sahel zone except in the north-east of Nigeria whilst a month early planting is expected in the south-coast of the country. On the other hand, the intervention may lead to a one-month delay in the planting of Cocoa and up to 3 months in the Guinea zone.

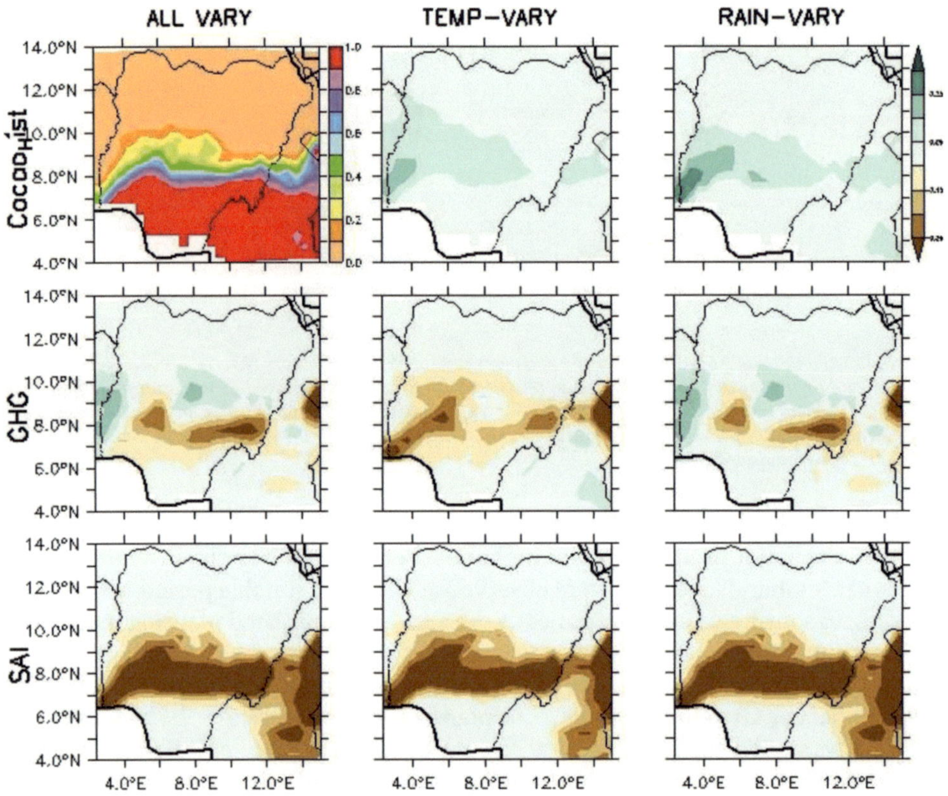

Figure 5.
The spatial distribution of Cocoas maize sensitivity to rainfall and temperature for the historical period over southern Africa (first panel, first row; 2011–2030) and relative changes over the historical period (first row, panel 2 & 3) and their projected future changes in the period (2070–2089) under the RCP8.5 scenario without (GHG) and with SAI (i.e., second and third rows, respectively). All-vary means both total monthly rainfall and monthly minimum and mean temperature varies, RAIN, monthly variation of minimum and mean temperature with annual mean total monthly rainfall (i.e., the same 12 monthly rainfall values are constant for the 30-year period) and TEMP; varying total monthly rainfall with constant annual minimum and mean temperature (i.e., the same 12 monthly minimum and mean temperature values are used for the 30-year period).

3.5 Sensitivity of Cocoa suitability to climate variables (rainfall and temperature)

We also examine the sensitivity of Cocoa suitability to temperature and rainfall to test independently the influence of their variation and trend for the three time periods, historical, GHG and SAI. In general, Cocoa suitability shows a decreasing suitability gradient from south to north over reference period for the all-vary experiment (see further description in Section 3.1) (**Figure 5**). The effect of rainfall variability at constant mean and minimum temperature (rain-vary) relative to the reference period (5a) shows no change in Cocoa suitability along the south coast, whilst southern savanna extending into the Sahel zone (5b-5i) will remain unsuitable. Similar suitability characteristics are also expected with variability in minimum and mean temperature at constant rainfall (temp-vary) over these zones. However, over the Guinea-savanna zones (7–10°N), about 0.15 suitability increase is expected and up to 0.25 along south-west boundary of Oyo and Ogun state.

The sensitivity of Cocoa suitability to rainfall and temperature variability shows a similar response under the GHG and SAI across the three experiments (all-vary, rain-vary and temp-vary) but with varied impact in the Guinea-savanna zones. Our results show the impact of GHG and deployment of SAI on all three experiments will lead to no change Cocoa suitability over south coast and northern Sahel zone of Nigeria. However, in the Guinea-savanna zone, the impact of the scenario across the three experiments varies for both GHG and SAI. The impact of GHG on all-vary and rain-vary experiments may lead to a SIV increase (0.15) in south-west boundary from Oyo to Ogun state and central savanna zone relative to the historical period. However, projected SIV decrease (up to 0.35) may be expected over the country except along south-west boundary from Oyo to Ogun state and central savanna zones. Also, SIV decrease (up to 0.35) is expected in the Guinea-savanna zones under temp-vary experiments. In contrast, the impact of SAI deployment on all three climate sensitivity over Nigeria is expected to lead to a decrease up to 0.35 in Cocoa suitability in the Guinea-savanna zones (6–10°N), whilst no change in suitability is expected in south coast (4–6°N) and northern Sahel zones (10–12°N) over Nigeria. This means that the deployment of SAI may lead to reduced area in the cultivation of Cocoa in Nigeria, as the south coast may be the only suitable for growing the crop.

In summary, we find that Cocoa suitability is sensitive to variability in both temperature and rainfall under SAI but more sensitive to rainfall variability with constant temperature under GHG. This agrees with past findings, by Challinor et al. [49] and Ramirez-Villegas et al. [37], that variability in rainfall affects the suitability of crops in Sub-Saharan Africa.

3.6 Percentage distribution of Cocoa suitability

We further examine the percentage distribution of different suitability condition to evaluate the impact of GHG and SAI on Cocoa at each grid point over Nigeria (**Figure 6**). The bar plot shows the percentage distribution of Cocoa suitability at each grid point over Nigeria for the historical period and the resultant impacts of GHG and SAI. Our result showed that about 42% of the land area in Nigeria is unsuitable for Cocoa cultivation and about 50% area is suitable (both suitable and highly suitable) over the historical period, whilst about 8% of the area is marginally suitable.

The impact of GHG and SAI on the percentage suitability distribution of Cocoa in Nigeria shows a similar pattern as the historical period albeit with varied magnitude. For example, global warming is projected to result in a decrease in suitable area for Cocoa cultivation in Nigeria relative to the historical period of about 18% and an increase (about 13%) in unsuitable area. This suggests that more areas will become unsuitable for growing Cocoa in Nigeria, as marginally suitable areas for Cocoa cultivation are projected to increase by about 2% across the country.

In addition, the impact of SAI shows a similar pattern in the percentage distribution of Cocoa suitability across grid points relative to the historical and GHG over Nigeria. SAI intervention may lead to a projected decrease of about 24% in suitable areas for Cocoa cultivation relative to the historical period and 18% decrease in comparison to the impact of GHG. In addition, the intervention may also lead to a further increase of about 14% and 24% in unsuitable areas for Cocoa relative to GHG and historical period, respectively, whilst no significant change was observed in areas with very marginal and marginal suitability for Cocoa in Nigeria.

The above result shows that GHG warming is projected to reduce the percentage of land suitable for Cocoa cultivation across Nigeria and SAI intervention would worsen this. Under both GHG and SAI scenarios, within in the GLENS modelling framework, there would be a resultant decrease in Cocoa production and yield in Nigeria [20].

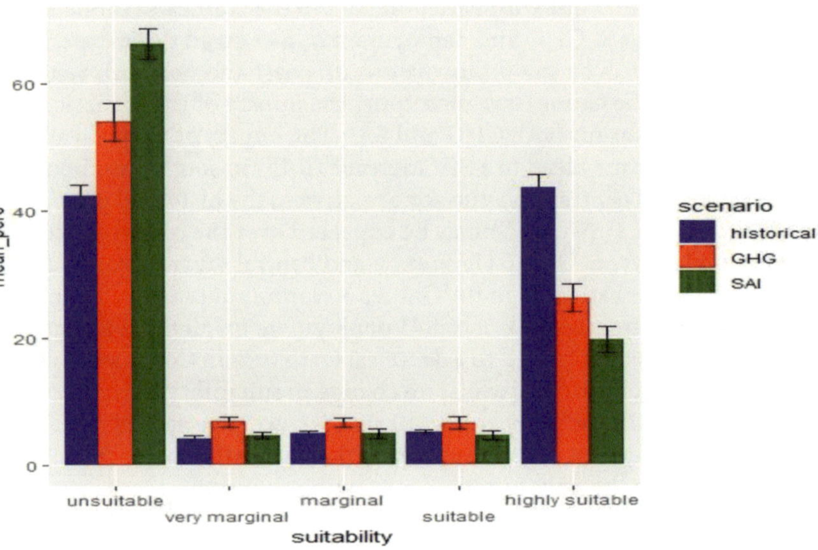

Figure 6.
A bar chart of percentage grid point distribution of Cocoa suitability over Nigeria for the historical period (2011–2030) and under the impact of GHG and SAI for the period (2070–2089).

3.7 Implication of SAI on Cocoa production and Nigerian economy

The availability of suitable land for cultivation is vital for crop growth and yield. The impact of SAI on Cocoa suitability and planting season is expected to have significant implication on socio-economy and GDP in Nigeria. The projected increase in unsuitable area in cultivating Cocoa and the corresponding decrease in suitable areas despite SAI intervention relative to the GHG impact over the country raises a concern considering the importance for Cocoa production to the economy of Nigeria. Cocoa is the major cash crops in Nigeria and most important agricultural export which can be processed into various products (e.g., Cocoa powder, Chocolates) for human consumption with a significant contribution to the socio-economy and GDP of the country via foreign exchange earnings [30]. SAI intervention on Cocoa suitability compared to the impact of climate change may further worsen the challenge of food and agricultural production in the region. This is because although SAI intervention reduces warming over Nigeria, its impact on Cocoa suitability relative to GHG will lead to a decrease in suitable areas for the Cocoa cultivation and may further lead to an increase in unsuitable areas relative to GHG as seen from our findings (**Figure 5**). The reduction of suitable land for Cocoa production and resultant decrease in production would lead to a decrease in Cocoa exports and its contribution to the GDP of the country [20, 30].

4. Conclusions

The present study examined the impact of SAI on Coca suitability and planting season over in Nigeria using the GLENS CESM1(WACCM) experiment. We used GLENS experiment, which is aimed at reducing the mean global surface temperature by injecting sulphur at four latitudes as a climate intervention measure to reduce the impact of greenhouse gases on Cocoa production in Nigeria. To examine the impact of

SAI intervention, we compared the GLENS output to historical and high-end emission scenario, RCP8.5 (GHG). We also examined the sensitivity of the crops to rainfall and temperature under GHG and SAI. The summary of our findings is listed below:

- Total monthly rainfall distribution decreases northward, with the least rainfall amount (80 mm) and the highest (240 mm) in the south coast in the Guinea zone over the historical period.

- The impact of GHG may lead to a 15 mm increase in total monthly rainfall by the end of century in south-west and north-east area of Nigeria with a decrease of the same magnitude in the other areas, whilst SAI intervention shows a reverse, leading to a decrease about 10 mm in total monthly rainfall and up to 20 mm in south-west Nigeria.

- GHG impact may lead to increase in both minimum and mean temperatures up to 3°C, whilst the SAI intervention will offset the warming resulting in a cooling over Nigeria.

- The impact of GHG may lead to an increase up to 0.2 in Cocoa suitability index value along the south west boundary of Ogun and Oyo with a decrease in the Guinea-savanna zones of Nigeria, whilst no change is expected in south coast (south of 6°N) and Sahel zone (north of 10°N).

- SAI intervention may lead to general decrease in suitable areas, whilst it results in a significant increase in unsuitable lands for Cocoa production in Nigeria.

- Percentage suitability distribution under GHG and SAI may result in a general decrease of 24% and 18%, respectively, in suitable available land with a corresponding increase about 14% and 24%, respectively, in unsuitable areas of Cocoa relative to the historical period over Nigeria.

- Cocoa is more sensitive to rainfall variability with a higher magnitude in the suitability index values and decrease in spatial extent over the region.

To our knowledge, this is the first study that examines the effect of SAI on Cocoa suitability in Sub-Saharan Africa. Hence, there are caveats to the interpretation of the result presented in this study. First, Ecocrop is a simple statistical model that evaluates crop suitability using total monthly rainfall and minimum and mean temperature climate variables. The model does not consider the effect of other factors, such as evapotranspiration and soil moisture, or non-climatic factors like pest and diseases and soil type which may further affect crop suitability. Also, the result is particular to solar geoengineering dataset, the GLENS experiment which used high-end representative scenario (RCP8.5) of climate geoengineering.

Nevertheless, the result from the study has helped improve our understanding of the potential impact of GHG and SAI on Cocoa suitability in Nigeria, the fourth largest producer of Cocoa in the world. Despite the caveats, one of advantage of using Ecocrop is that it is a simple and straight forward crop model to use with limited data requirement. Moreover, the observation and model representation of climate variables such as rainfall and temperature provide a basis for confidence in their outputs and the results are consistent with previous study on climate change and geoengineering impacts with complex models [20, 37]. Further studies on the impact of SAI on other crop types, such as legumes, root and tuber, other horticultural crops including vegetables are recommended in the quest to understand food security risk under both GHG emissions and SAI. These studies should use more than just the GLENS datasets and models should be used to provide robust information on the impact of SAI on Cocoa.

Acknowledgements

We acknowledge the financial support of the DEGREES Modelling Fund (DMF) of the DEGREES Initiative. Further supports were received from the National Research Fund (NRF) Scarce Skills Postdoctoral Fellowship, *ACCESS Annual Cycle* and *Seasonality Project* (ACyS); the DAAD within the framework of the ClimapAfrica programme with funds from the German Federal Ministry of Education and Research. We also appreciate the Centre for High Performance Computing (CHPC), South Africa for the computational resources. The data used in this study are freely available to the community via the Earth System Grid (https://www.cesm.ucar.edu/projects/community-projects/GLENS/). The CESM project is supported primarily by the National Science Foundation.

Author details

Temitope S. Egbebiyi[1,2*], Christopher Lennard[1], Pinto Izidine[3], Romaric C. Odoulami[2], Piotr Wiolski[1] and Akintunde I. Makinde[1]

1 Department of Environmental and Geographical Science, Climate Systems Analysis Group, University of Cape Town, Cape Town, South Africa

2 African Climate and Development Initiative, University of Cape Town, Cape Town, South Africa

3 KNMI - Royal Netherlands Meteorological Institute, Netherlands

*Address all correspondence to: temitope.egbebiyi@uct.ac.za

References

[1] Afolayan OS. Cocoa production pattern in Nigeria: The missing link in regional Agro-economic development. Analele Universității din Oradea, Seria Geografie. 2020;**30**(1):88-96. DOI: 10.30892/auog.301110-815

[2] Edeki SO, Adeosun ME, Akinlabi GO, Ofuyatan OM. Datasets for correlation dynamics of cocoa production in South Western Nigeria. 2018. DOI: 10.1016/J.DIB.2018.03.076

[3] Wessel M, Quist-Wessel PF. Cocoa production in West Africa, a review and analysis of recent developments. NJAS: Wageningen Journal of Life Sciences. 2015;**74**(1):1-7

[4] Kehinde A. Agricultural cooperatives and improved technologies adoption among smallholder farmers in cocoa-based farming systems of southwestern Nigeria. International Journal of Agricultural Management and Development. 2021;**11**(4):467-483

[5] Kehinde AD. Access to trade credit and its impact on the use of European Union (EU) approved pesticides among smallholder cocoa farmers in Ondo state, Nigeria. Heliyon. 2022;**8**(12)

[6] Nkamleu GB, Nyemeck J, Gockowski J. Technology Gap and Efficiency in Cocoa Production in West and Central Africa: Implications for Cocoa Sector Development. African Development Bank: Abidjan, Côte d'Ivoire; 2010

[7] Agbongiarhuoyi AE, Abdulkarim IF, Fawole OP, Obatolu BO, Famuyiwa BS, Oloyede AA. Analysis of farmers' adaptation strategies to climate change in cocoa production in Kwara state. Journal of Agricultural Extension. 2013;**17**(1):10-22

[8] Olayioye A, Olaniran OA, Olaifa JI. Effect of powdered castor oil seed (Ricinus communis L.) on some internal organs of albino rat. International Journal of Applied Agriculture and Apiculture Research. 2014;**10**(1-2):98-111

[9] Oyekale AS. Impact of climate change on cocoa agriculture and technical efficiency of cocoa farmers in south-West Nigeria. Journal of Human Ecology. 2012;**40**(2):143-148

[10] Adebiyi S, Uwagboe EO, Agbongiarhuoyi AE, Oluyole KA, Famuyiwa BS, Abdulkarim IF, et al. Climate change and adaptation strategies on cocoa production in Ibarapa central of oyo state, Nigeria. International Journal of Contemporary Research and Review. 2018;**9**:20478-20483. DOI: 10.15520/ijcrr/2018/9/07/547

[11] Omotayo FS, Oguntunde PG, Olufayo AA. Influence of climatic variables on whole-plant water use of cocoa under limited soil moisture condition. Journal of Agriculture and Ecology Research International. 2019;**19**(4):1-8

[12] Ojo AD, Sadiq I. Effect of climate change on cocoa yield: A case of cocoa research institute (CRIN) farm, Oluyole local government Ibadan Oyo state. Journal of Sustainable Development in Africa. 2010;**12**(1):350-358

[13] Black E, Pinnington E, Wainwright C, Lahive F, Quaife T, Allan RP, et al. Cocoa plant productivity in West Africa under climate change: A modelling and experimental study. Environmental Research Letters. 2020;**16**(1):014009. DOI: 10.1088/1748-9326/abc3f3

[14] Nwachukwu IN, Ezeh CI, Emerole CO. Effect of climate change on

cocoa productivity in Nigeria. African Crop Science Journal. 2012;**20**:487-491

[15] Jamal AM, Antwi-Agyei P, Baffour-Ata F, Nkiaka E, Antwi K, Gbordzor A. Gendered perceptions and adaptation practices of smallholder cocoa farmers to climate variability in the central region of Ghana. Environmental Challenges. 2021;**5**:100293

[16] Wigley TM. A combined mitigation/geoengineering approach to climate stabilization. Science. 2006;**314**(5798):452-454

[17] Kravitz B, MacMartin DG, Wang H, Rasch PJ. Geoengineering as a design problem. Earth System Dynamics. 2016;7(2):469-497. DOI: 10.5194/esd-7-469-2016

[18] Robock A. Stratospheric aerosol geoengineering. AIP Conference Proceedings. 2015;**1652**:183-197

[19] MacMartin DG, Kravitz B, Tilmes S, Richter JH, Mills MJ, Lamarque J-F, et al. The climate response to stratospheric aerosol geoengineering can be tailored using multiple injection locations. Journal of Geophysical Research. 2017;**122**:12574-12590

[20] Xia L, Robock A, Cole J, Curry CL, Ji D, Jones A, et al. Solar radiation management impacts on agriculture in China: A case study in the geoengineering model intercomparison project (GeoMIP). Journal of Geophysical Research: Atmospheres. 2014;**119**(14):8695-8711

[21] Cheng W, MacMartin DG, Dagon K, Kravitz B, Tilmes S, Richter JH, et al. Soil moisture and other hydrological changes in a stratospheric aerosol geoengineering large ensemble. Journal of Geophysical Research: Atmospheres. Blackwell Publishing Ltd. 2019;**124**(23):12773-12793. DOI: 10.1029/2018JD030237

[22] Pinto I, Jack C, Lennard C, Tilmes S, Odoulami RC. Africa's climate response to solar radiation management with stratospheric aerosol. Geophysical Research Letters. 2020;**47**(2). DOI: 10.1029/2019GL086047

[23] Abiodun BJ, Odoulami RC, Sawadogo W, Oloniyo OA, Abatan AA, New M, et al. Potential impacts of stratospheric aerosol injection on drought risk managements over major river basins in Africa. Climatic Change. 2021;**169**:1-19. DOI: 10.1007/s10584-021-03268-w

[24] Simpson IR, Tilmes S, Richter JH, Kravitz B, MacMartin DG, Mills MJ, et al. The regional hydroclimate response to stratospheric sulfate geoengineering and the role of stratospheric heating. Journal of Geophysical Research: Atmospheres. 2019;**124**(23):12587-12616. DOI: 10.1029/2019JD031093

[25] Proctor J, Hsiang S, Burney J, et al. Estimating global agricultural effects of geoengineering using volcanic eruptions. Nature. 2018;**560**:480-483. DOI: 10.1038/s41586-018-0417-3

[26] Abiodun BJ, Lawal KA, Salami AT, Abatan AA. Potential influences of global warming on future climate and extreme events in Nigeria. Regional Environmental Change. 2013;**13**:477-491. DOI: 10.1007/s10113-012-0381-7

[27] Abiodun BJ, Pal JS, Afiesimama EA, Gutowski WJ, Adedoyin A. Simulation of west African monsoon using RegCM3 part II: Impacts of deforestation and desertification. Theoretical and Applied Climatology. 2008;**93**:245-261. DOI: 10.1007/s00704-007-0333-1

[28] Gershon O, Mbajekwe C. Investigating the nexus of climate change

and agricultural production in Nigeria. International Journal of Energy Economics and Policy. 2020;**10**(6):1-8. DOI: 10.32479/ijeep.9843

[29] OECD/FAO. Agriculture in Sub-Saharan Africa: Prospects and challenges for the next decade. OECD-FAO Agricultural Outlook 2016-2025. Paris, France: OECD Publishing; 2016. pp. 59-95. Available from: https://stats.oecd.org/Index.aspx?DataSetCode=HIGH_AGLINK_2016

[30] Schroth G, Läderach P, Martinez-Valle AI, Bunn C, Jassogne L. Vulnerability to climate change of cocoa in West Africa: Patterns, opportunities and limits to adaptation. Science of the Total Environment. 2016;**556**:231-241. DOI: 10.1016/j.scitotenv.2016.03.024

[31] Tilmes S, Richter JH, Kravitz B, MacMartin DG, Mills MJ, Simpson IR, et al. CESM1(WACCM) stratospheric aerosol geoengineering large ensemble project, B. American Meteorological Society. 2018;**99**:2361-2371. DOI: 10.1175/BAMS-D-17-0267.1

[32] Mills MJ, Richter JH, Tilmes S, Kravitz B, MacMartin DG, Glanville AA, et al. Radiative and chemical response to interactive stratospheric sulfate aerosols in fully coupled CESM1(WACCM). Journal of Geophysical Research-Atmospheres. 2017;**122**:13061-13078. DOI: 10.1002/2017JD027006

[33] Odoulami RC, New M, Wolski P, Guillemet G, Pinto I, Lennard C, et al. Stratospheric aerosol geoengineering could lower future risk of 'Day Zero'level droughts in Cape Town. Environmental Research Letters. 2020;**15**(12):124007. DOI: 10.1088/1748-9326/abbf13

[34] Harris I, Jones PD, Osborn TJ, Lister DH. Updated high-resolution grids of monthly climatic observations - The CRU TS3.10 dataset. International Journal of Climatology. 2014;**34**(3):623-642. DOI: 10.1002/joc.3711

[35] Cannon AJ, Sobie SR, Murdock TQ. Bias correction of GCM precipitation by quantile mapping: How well do methods preserve changes in quantiles and extremes? Journal of Climate. 2015;**28**(17):6938-6959

[36] Famien AM, Janicot S, Delfin Ochou A, Vrac M, Defrance D, Sultan B, et al. A bias-corrected CMIP5 dataset for Africa using the CDF-t method - A contribution to agricultural impact studies. Earth System Dynamics. 2018;**9**(1):313-338. DOI: 10.5194/esd-9-313-2018

[37] Ramirez-Villegas J, Jarvis A, Läderach P. Empirical approaches for assessing impacts of climate change on agriculture: The EcoCrop model and a case study with grain sorghum. Agricultural and Forest Meteorology. 2013;**170**:67-78. DOI: 10.1016/j.agrformet.2011.09.005

[38] Hijmans RJ, Guarino L, Cruz M, Rojas E. Computer tools for spatial analysis of plant genetic resources data: 1. DIVA-GIS. Plant Genetic Resources Newsletter. 2001;**127**:15-19

[39] Egbebiyi TS, Crespo O, Lennard C, Zaroug M, Nikulin G, Harris I, et al. Investigating the potential impact of 1.5, 2 & 3°C global warming levels on crop suitability and planting season over West Africa. PeerJ-Life & Environment Journal. 2020;**8**:e8851. DOI: 10.7717/peerj.8851

[40] Egbebiyi TS, Lennard C, Crespo O, Mukwenha P, Lawal S, Quagraine K. Assessing future Spatio-temporal changes in crop suitability and planting season over West Africa: Using the concept of crop-climate departure.

Climate. 2019;**7**(9):1-30. DOI: 10.3390/cli7090102

[41] Ramírez-Villegas J, Lau C, Köhler A-K, Signer J, Jarvis A, Arnell N, et al. Climate Analogues: Finding Tomorrow's Agriculture Today CGIAR Research Program on Climate Change, Agriculture and Food Security (CCAFS). 2011. Available from: www.ccafs.cgiar.org

[42] Klutse NAB, Ajayi VO, Gbobaniyi EO, Egbebiyi TS, Kouadio K, Nkrumah F, et al. Potential impact of 1.5 °c and 2 °c global warming on consecutive dry and wet days over West Africa. Environmental Research Letters. 2018;**13**(5). DOI: 10.1088/1748-9326/aab37b

[43] Maúre G, Ndebele-Murisa M, Nikulin G, Lennard C, Meque A, Muthige M, et al. The southern African climate under 1.5 °C and 2 °C of global warming as simulated by CORDEX regional climate models. Environmental Research Letters. 2018;**13**(6):065002. DOI: 10.1088/1748-9326/aab190

[44] Patel TD, Odoulami RC, Pinto I, Egbebiyi TS, Lennard C, Abiodun BJ, et al. Potential impact of stratospheric aerosol geoengineering on projected temperature and precipitation extremes in South Africa. Environmental Research: Climate. 2023;**2**(3):035004

[45] Nikulin G, Lennard C, Dosio A, Kjellström E, Chen Y, Hänsler A, et al. The effects of 1.5 and 2 degrees of global warming on Africa in the CORDEX ensemble. Environmental Research Letters. 2018;**13**(6):065003. DOI: 10.1088/1748-9326/aab1b1

[46] Irvine PJ, Keith DW. Halving warming with stratospheric aerosol geoengineering moderates policy-relevant climate hazards. Environmental Research Letters. 2020;**15**(4):044011

[47] Läderach P, Martinez-Valle A, Schroth G, Castro N. Predicting the future climatic suitability for cocoa farming of the world's leading producer countries, Ghana and Côte d'Ivoire. Climatic Change. 2013;**119**(3-4):841-854

[48] Ofori-Boateng K, Insah B. The impact of climate change on cocoa production in West Africa. International Journal of Climate Change Strategies and Management. 2014;**6**(3):296-314

[49] Challinor AJ, Watson J, Lobell DB, Howden SM, Smith DR, Chhetri N. A meta-analysis of crop yield under climate change and adaptation. Nature Climate Change. 2014;**4**(4):287-291. DOI: 10.1038/nclimate2153

Chapter 3

Climate Variability and Outlook of Cocoa Production in Côte D'ivoire under Future Climate

Antoine Alban Kacou M'bo, Mamadou Cherif, Kouakou Kouadio, Mahyao Germain Adolphe, Adama Bamba, Evelyne N'Datchoh Toure, Alla Kouadio Okou, Renée Brunelle, Yanick Rouseau and Daouda Koné

Abstract

Cocoa supports about 3.5 million people. Farmers produce each year 1.5 million ton. This performance hides production constraints, the most is climate variability. The climatic variables, temperature, precipitation, and 16 climatic indices were identified to assess the potential impacts on cacao in the past year, currently and under future climate. The climate data in the southern and central cocoa production zone were analysed for periods of 2021–2050 and 2041–2070. The climate reference period is 1981–2010. The climate projections are from the CORDEX RCP 4.5 and 8.5. The results suggest an increase in daily temperature of 1.0–2.1°C in the central region and 0.9–2.0°C in the southern region by 2041–2070. Cocoa could be affected by the projected changes, especially in the central region where the maximum daily temperature at which production is reduced (33°C) would be exceeded between 92 and 142 days per year by this time horizon. The direction of changes in precipitation cannot be established due to a lack of consensus between the climate models analysed. However, the little rainy season would start slightly earlier, potentially reducing the duration of the little dry season between the rainy seasons. The climate scenarios enhanced deterioration of growing environment conditions. It is necessary to take adaptation measures to mitigate climate impacts.

Keywords: climate, cacao, production, sustainability, Côte d'Ivoire

1. Introduction

Cocoa is one of Côte d'Ivoire's main sources of foreign currency and the driving force behind its economic growth. It alone accounts for 15% of GDP. With an annual production of more than one million five hundred thousand tons since 2015 [1], this performance is due not only to the valuable results of agronomic research but also to the favourable ecological conditions in the southern half of the country. However, this spectacular overall result, which has made Côte d'Ivoire the world's leading producer, is based on a rapid and poorly controlled increase in the area under cultivation, leading to extensive deforestation. Furthermore, the ageing of cocoa trees and the lack of appropriate maintenance have led to a drop in the productivity of orchards, resulting in the impoverishment of communities [2].

Cocoa production is subject to significant inter-annual variability, accentuated by the action of soil and climate hazards, strong parasitic pressure from insects, and diseases such as *Swollen shoot* [3, 4] and the brown rot disease caused by *Phytophtora sp*., [5]. The greatest production constraint since 2010 has been climatic variability [6–8]. This can be seen in the significant reduction in the total leaf area of cocoa trees, and in the fall of flowers and young fruit [9–11], leading to a drastic drop in yield and the failure to rehabilitate and replant cocoa orchards. The effects of climate variability can also be seen in the interruption of young fruit growth, the reduction in bean and pod size, and the deterioration in bean quality [12–14].

Cocoa production areas experience significant inter-annual variability [15]. Climate is one of the main factors explaining this variability. It is important to note that rainfall is the most significant factor in cocoa growing, as a prolonged lack of water during the flowering phase can lead to a drop in production. Other climatic (temperature, humidity, solar radiation), ecological, biological, and physical factors can also have a significant influence on the phenology of the cocoa tree and its yields. The cocoa tree, an ombrophilous plant, requires specific climatic conditions for its development [16], annual rainfall of between 1300 and 2000 mm with a limited number of dry days (less than 3 months), an average daily temperature of between 24°C and 28°C, relative humidity of between 80% and 90%, daily sunshine of more than four hours, and deep, well-aerated soil rich in clay and humus. To develop and improve cocoa-growing conditions in Côte d'Ivoire, agrometeorological studies would contribute to a better understanding of the climatic factors likely to influence the plant's environment. This could not only help to improve farming practices, but also assess certain risks to the cocoa tree due to irregular rainfall, extreme temperatures and humidity, and low insolation. The effects of climate shock are expected to vary from one region to another, implying a need for adaptation or mitigation by context [8, 17]. Communities dependent on cocoa production are increasingly vulnerable to the impacts of climate change. These impacts stem from an increase in temperature and a change in rainfall patterns.

To better assess the impact of the climate, climate scenarios are an important step in assessing the vulnerabilities and impacts of climate change, and in identifying adaptation strategies capable of sustaining cocoa production over the coming decades. A climate scenario is a plausible description of the future state of the climate [18]. According to the Ouranos report [19], climate scenarios are produced by combining in-situ observations used as a reference dataset and climate projections for a given climate variable. This combination results in a climate scenario or a sequence of values associated with this variable for a period extending over several decades and for a given frequency. Climate models refer to greenhouse gas (GHG) emission scenarios as Representative Concentration Pathways (RCPs) to represent future radiative forcing. There are several groups of RCP models, but the most widely used are RCP 4.5 and RCP 8.5, which correspond respectively to a decrease in GHG emissions (optimistic scenario) and a constant increase in emissions throughout the century (pessimistic scenario). These models are part of the CORDEX (COordinated Regional Climate Downscaling EXperiment) – Africa domain [20–22]. In the present analysis, the RCP 4.5 and RCP 8.5 models were used to develop climate scenarios for the cocoa-growing area in Côte d'Ivoire. The projections are likely values that lie within a confidence interval. 16 climatic indices were defined according to the requirements of the cocoa tree, and calculations were then made taking into account the uncertainty associated with the inter-model differences between the RCP 4.5 and RCP 8.5 greenhouse gas (GHG) emission scenarios. These calculations were used to develop climate scenarios for the cocoa-growing zone in Côte d'Ivoire, to interpret the impact of fluctuations in climatic variables on the cocoa tree in future decades, and to identify endogenous adaptation practices to cope with the future climate.

2. Zones and time horizons

The cocoa production zone in Côte d'Ivoire is divided into two climatic regions. The central region (between 5.5° and 8° north latitude) and the southern region (below 5.5° north latitude) (**Figure 1**). In each of the two climate regions, the projected climate parameters are temperature and rainfall over two (2) time horizons (2021 to 2050 and 2041 to 2070) compared with the reference period 1981–2010. The study began in 2019, and as the data for the 2011–2020 decade is not complete, the 1991–2020 normal has not been used as a reference. The length of the time horizons of thirty (30) years follows the normal standard in climatology [23, 24]. This duration is generally long enough to obtain representative climate statistics, except for extreme and scarcer events [19]. A total of 16 climate indices were identified and calculated to assess the impact of climate change on cocoa production. The parameters and thresholds associated with these indices are based on information presented in scientific articles and on WASCAL-CEA-CCBAD experts' knowledge of cocoa physiology and climate. These indices depend on two climatic variables, either temperature, rainfall, or both. The interpretation of the results is grouped by climatic variable.

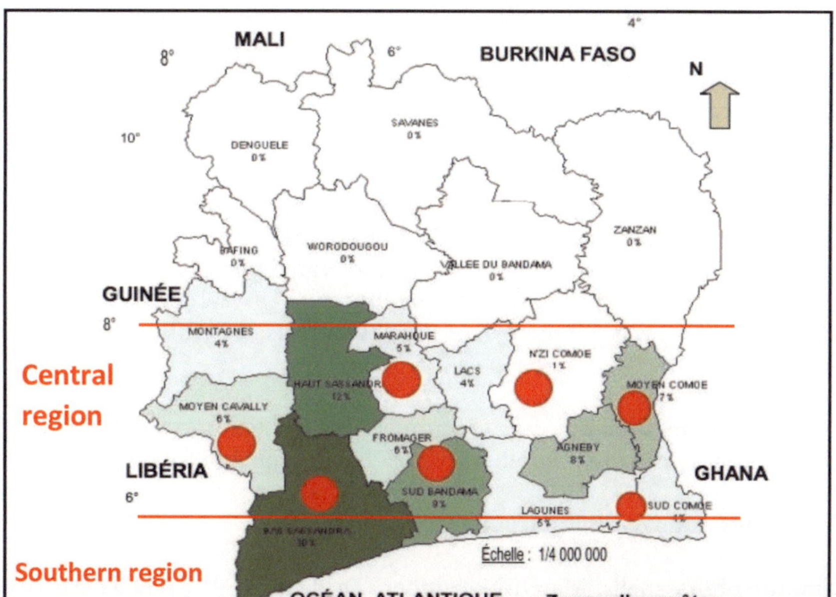

Figure 1.
Ivorian cocoa production area is divided into climatic regions.

2.1 Temperature

The ideal conditions to grow cocoa are when the annual average daily maximum temperature is between 30°C and 32°C and the annual average daily minimum temperature is between 18°C and 21°C [12]. These conditions are associated with maximum photosynthesis. A monthly average daily minimum of at least 15°C is necessary for plant health. A reduction in the growth and development of cocoa trees can occur if the number of days with a maximum temperature of over 33°C is too great [25, 26]. Taking into account this information on temperature-related aspects from the literature review, the following climatic variables were selected: average daily temperature (tas, °C), minimum daily temperature (tasmin, °C), and maximum daily temperature (tasmax, °C). These were used to calculate the following climatic indices:

- Average daily minimum temperature (TNg, °C),
- Mean maximum daily temperature (TXg, °C),
- Number of months per year with a low daily minimum temperature (TNgMonthsBelow) This is the number of months per year when the average value of the daily minimum temperatures (tasmin) is below a lower limit value for the cocoa tree. The threshold selected is 15°C.
- Number of days per year with a high daily maximum temperature (TXDaysAbove) This is the number of days per year when the daily maximum temperature (tasmax) is greater than or equal to 33°C.

2.2 Rainfall

Rainfall between 1500 and 2000 mm per year is generally considered to be the most favourable for cocoa farming. Too much rain can increase the occurrence of diseases and attract pests, thus increasing the mortality rate of cocoa trees. Less than 1200 mm/year can lead to reduced root growth, leaf drop, and reduced plant growth. Cocoa trees must receive at least 700 mm of rain during the rainy season. What's more, for cocoa to ripen properly, the rainy season must last 4 consecutive months, from the flowering phase to the end of the main harvesting season (March to November). In addition, dry periods of more than 14 days can lead to a drop in production, and a dry season of more than 3 months is not tolerated by cocoa trees. Taking into account the information on rainfall from the literature review, the Daily rainfall (pr, mm) climate variable was adopted. This variable was used to calculate the following climate indices:

- Number of rainy days per year (Rnnmm) This is the number of days per year on which the sum of daily rainfall is greater than or equal to 5 mm.
- Cumulative annual rainfall (PrcpTot; mm)
- Duration of dry spells, March-November (DryDurTot_1; days) This is the total duration of dry spells between 1 March (day 60) and 30 November (day 334). A dry spell is defined as a sequence of at least 14 days in which the cumulative daily rainfall is less than 1 mm.

- Duration of dry spells, November-March (DryDurTot_2; days): This is the total duration of dry spells, between 1st November (day 305) and 31st March (day 90). A dry period is defined as a period during which rainfall is less than 70 mm per period of 30 consecutive days. The pr variable was also used to calculate the climatic indices associated with the main (first) rainy season:

- Start of the main rainy season (RainStart_1; day of year) This is the day (value from 1 to 365) on which the main rainy season begins. This day normally occurs after 11 March. The algorithm identifies the first day after this date (day 71 of the year or more) on which a total of 20 mm of rain is received in two days without there being a dry period lasting 7 days in the 30 days following the first two rainy days. A dry day corresponds to a daily rainfall total of less than 1 mm.

- End of the main rainy season (RainEnd_1; day of the year) This is the day (value from 1 to 365) on which the main rainy season ends. This day normally occurs after 11 July. The algorithm identifies the first day after this date (day 193 of the year or more) on which daily rainfall is less than 10 mm/day for 10 consecutive days. The season cannot end before it has begun, so this index depends on RainStart_1. If no days are detected for a year, then that year is excluded from the calculation for the climate scenario being analysed. Also, the RainEnd_1 index depends on the RainStart_2 index, which means that the indices associated with the short rainy season are calculated before those for the long rainy season.

- Duration of the main rainy season (RainDur_1; days) This is the duration of the main rainy season, i.e., the number of days between the start and end of the rainy season, plus one day.

- Cumulative rainfall during the main rainy season (RainQty_1; mm) The pr variable was also used to calculate climate indices for the short (second) rainy season:

- Start of little rainy season (RainStart_2; day of year) This is the day (value from 1 to 365) on which the short rainy season begins. This day normally occurs after 11 August. The algorithm identifies the first day after this date (day 223 of the year or more) on which a total of 20 mm of rain is received in two days without there being a dry period lasting 7 days in the 30 days following the first two rainy days. A dry day corresponds to a daily rainfall total of less than 1 mm.

- End of short rainy season (RainEnd_2; days of the year) This is the day (value from 1 to 365) on which the short rainy season ends. This day normally occurs after 1st November. The algorithm tries to identify the first day after this date (day 306 of the year or more) on which daily rainfall is less than 10 mm/day for 10 consecutive days. The season cannot end before it has begun, so this index depends on RainStart_2. If no days are detected for a year, then that year is excluded from the calculation for the climate scenario being analysed.

- Duration of short rainy season (RainDur_2; days) This is the duration of the short rainy season, i.e., the number of days between the start and end of the rainy season, plus one day.

- Cumulative rainfall during the short rainy season (RainQty_2; mm).

3. Climate variability in the cocoa-growing zone

3.1 Current climate in the centre region

3.1.1 Temperature

Temperatures are high in February to March (**Figure 2**). During these months, the average temperature is 27.1°C, while the average maximum daily temperature is 32.6° C. The greatest inter-annual variation occurs between December and March. Temperatures are lowest in July and August. During these months, the average temperature is 24.1°C, while the average maximum daily temperature is 27.8°C. Temperatures rise slightly in October-November. The lowest temperatures are recorded in February–March (23.3°C) and October-November (22.1°C). In January, July, August, and December, the average daily minimum temperature is 21.7°C.

3.1.2 Precipitation

The sources of information available to assess the projected changes in these indices were: (1) rainfall climate scenarios, which were produced from reanalyses (the ERA5-Land ensemble) and climate projections (the CORDEX ensemble) using the

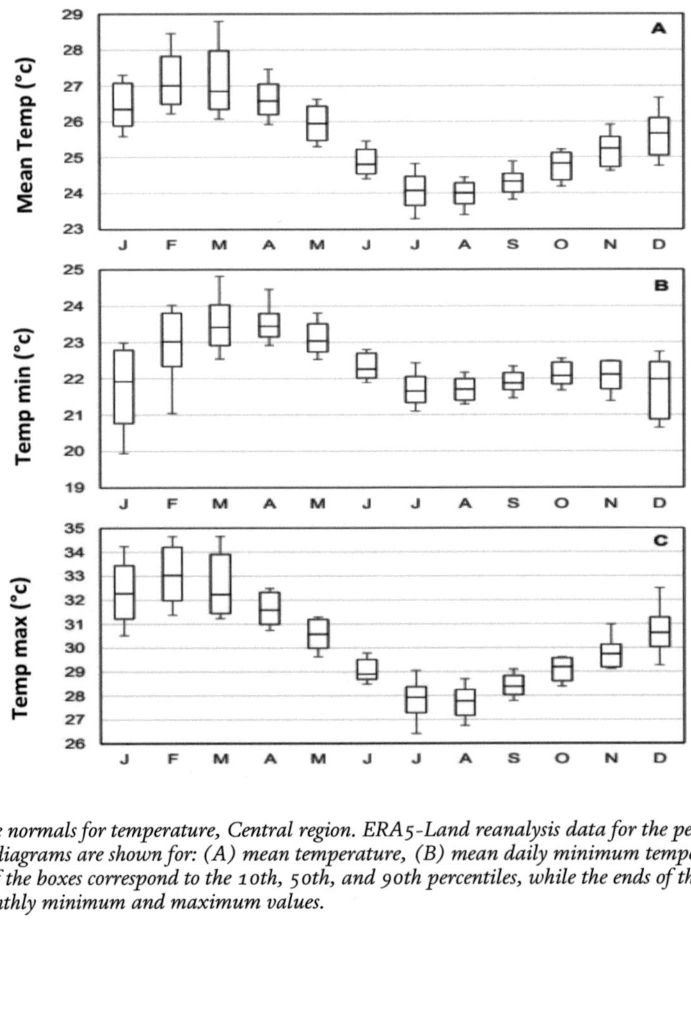

Figure 2.
Monthly climate normals for temperature, Central region. ERA5-Land reanalysis data for the period 1981–2010 were used. The diagrams are shown for: (A) mean temperature, (B) mean daily minimum temperature The vertical limits of the boxes correspond to the 10th, 50th, and 90th percentiles, while the ends of the error bars indicate the monthly minimum and maximum values.

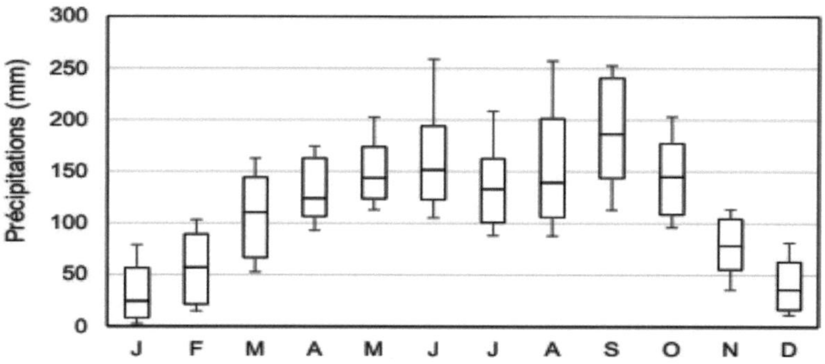

Figure 3.
Monthly climate normals for precipitation, Central region. ERA5-Land reanalysis data for the period 1981–2010 were used. The vertical limits of the boxes correspond to the 10th, 50th, and 90th percentiles, while the ends of the error bars indicate the minimum and maximum monthly values.

bias adjustment method, (2) rainy season detection parameters, and (3) the usual length of the break between rainy seasons in Côte d'Ivoire. An analysis of rainfall climate scenarios was completed to estimate monthly accumulations for the reference period and to visualize the characteristics of the two seasons, in terms of start, end, duration, and accumulation.

The data indicate more abundant rainfall between March and October inclusive, as well as the presence of two rainy seasons in the central region. Monthly rainfall is highest in September, followed by June, according to the median values (**Figure 3**). Interannual variability is greatest between June and September. The parameters initially chosen to detect the end of the rainy seasons during the reference period result in a long rainy season that ends 8 days before the start of the short season, on average (**Table 1**); this implies a quasi-unimodal distribution of rainfall. The period during which there is a decrease in rainfall between seasons seems too short considering the usual length of this period in Côte d'Ivoire (4–6 weeks) (**Figure 4**). Also, the timing of the break does not correspond to the decrease in rainfall between days ~200 and ~230 for the reference period.

Season	Indice	Initial parameters (<5 mm/d × 20d)		Revised parameters (<10 mm/d × 10d)	
Major rainy season	Start	07 April	Day 97	7 April	Day 97
	End	19 April	Day 231	31 July	Day 212
	Duration	135 Days		116 Days	
	Precipitations	606 mm (46% of total annual)		524 mm (39% of total annual)	
Short rainy season	Start	27 August	Day 239	27 August	Day 97
	End	04 December	Day 338	17 November	Day 212
	Duration	100 Days		83 Days	
	Precipitations	430 mm (32% of total annual)		392 mm (30% of total annual)	
Total	Duration	235 Days		200 Days	
	Precipitations	1036 mm (78% of total annual)		916 mm (69% of total annual)	

Table 1.
Characteristics of rainy seasons detected in the central region.

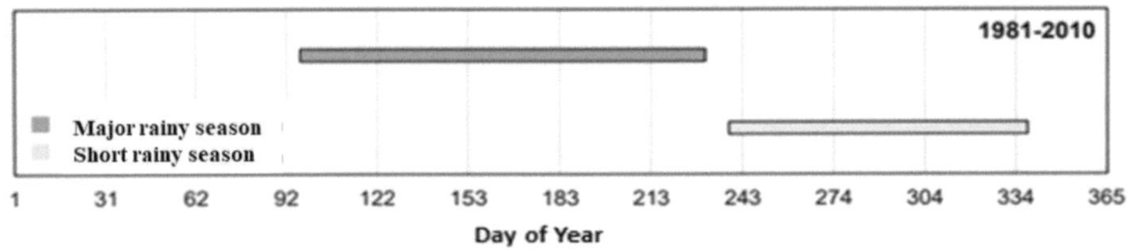

Figure 4.
Rainy seasons, based on initial definition of ONSET and cessation of rainy season for the central region.

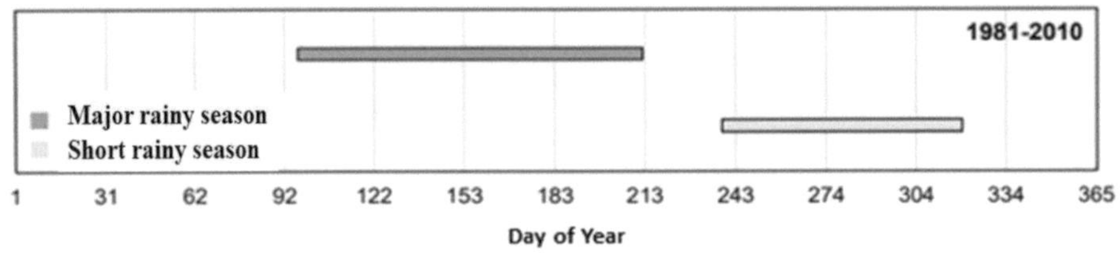

Figure 5.
Rainy seasons, based on revised definition of ONSET and cessation of rainy season for the central region.

These preoccupations prompted a sensitivity analysis of the parameters of the end-of-rainy-season indices. The formulation of these indices is that the season ends after the cumulative daily rainfall has been below a threshold P (mm/day) for a number of consecutive days t. A total of 25 combinations of thresholds P and t were tested with the reanalysis data for the reference period. For each of the combinations of these thresholds, the end dates of the rainy seasons were noted (averaged over the region). One of the combinations tested resulted in a dry period lasting 24 days (**Table 1**), which seems more compatible with reality (**Figure 5**). Using these rainy season detection parameters, the main rainy season extends from 7 April to 31 July, on average (**Table 1**). This corresponds to duration of 116 days. The short rainy season extends from 27 August to 17 November, lasting 83 days. Note that we now detect a dry season between the two rainy seasons with the revised parameters.

3.2 Current climate in the southern region

3.2.1 Temperature

The 30-year (1981–2010) monthly averages of mean, minimum, and maximum daily temperatures show important intra-annual variability. The high values of temperatures are recorded between February and April with average values about 23.6°C, 26.1°C and 30.2°C. The lower temperatures are observed in July and August during the little dry season. During these months, the average values are around 22.1°C for the minimum temperature, 23.9°C for the mean temperature and 26.7°C for the maximum temperature (**Figure 6**).

3.2.2 Precipitation

The projected changes in precipitation were assessed using (i) rainfall climate scenarios, which were produced from reanalyses (the ERA5-Land ensemble) and bias corrected climate projections (the CORDEX ensemble), (ii) existing definition of

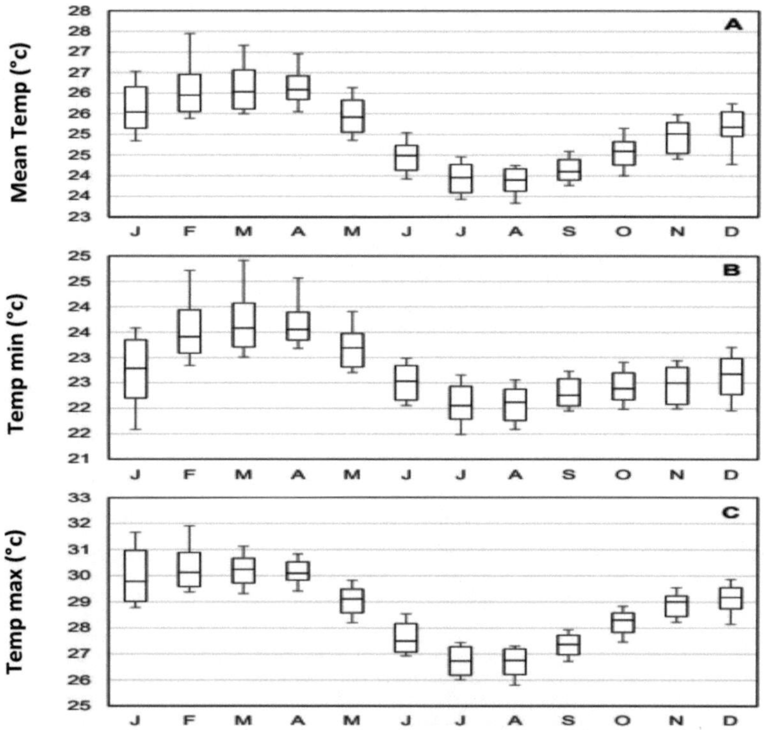

Figure 6.
Monthly climate normals for temperature, southern region ERA5-Land reanalysis data for the period 1981–2010 were used. The diagrams are shown for: (A) mean temperature, (B) minimum temperature, (C) and daily maximum temperature. The vertical limits of the boxes correspond to the 10th, 50th, and 90th percentiles, while the ends of the error bars indicate the minimum and maximum monthly values.

ONSET and cessation of rainy seasons, and (iii) usual duration of the little dry rainy season in the considered areas of Côte d'Ivoire. The seasonal analysis of the rainfall shows two rainy seasons in the southern region as well. Monthly rainfall is highest in May to June, followed by September-October, according to the median values (**Figure 7**). Interannual variability is greatest between June and August. The parameters initially chosen to detect the end of the rainy seasons result in a long

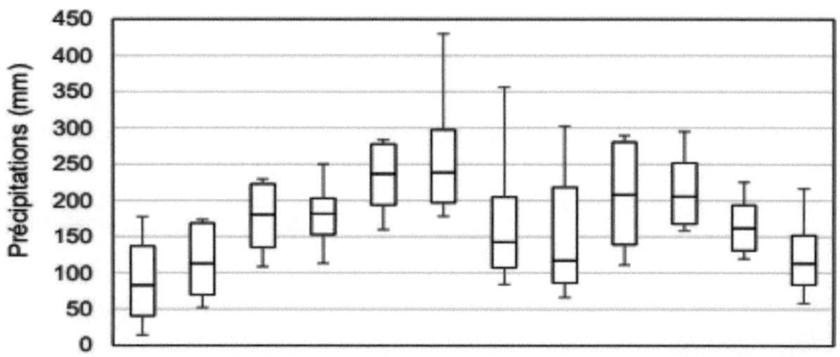

Figure 7.
Monthly climate normals for precipitation, southern region. ERA5-Land reanalysis data for the period 1981–2010 were used. The vertical limits of the boxes correspond to the 10th, 50th, and 90th percentiles, while the ends of the error bars indicate the minimum and maximum monthly values.

rainy season that ends 16 days before the start of the short season, on average (**Table 2**); as was the case in the central region, this implies a quasi-unimodal distribution of rainfall. The period during which there is a decrease in rainfall between seasons seems short, considering the usual length of this period in Côte d'Ivoire (4–6 weeks) (**Figure 8**). Also, the timing of the break does not correspond to the decrease in rainfall between days ~195 and ~235 for the reference period (**Figure 9**).

A sensitivity analysis was also carried out for the southern region. One of the combinations tested (10 mm/day for 10 days) resulted in a dry period lasting 31 days (**Table 2**) between the two rainy seasons, which seems more compatible with the reality in Côte d'Ivoire (**Figure 10**).

Season	Indice	Initial parameters (<5 mm/d × 20d)		Revised parameters (<10 mm/d × 10d)	
Major rainy season	Start	28 March	Day 87	28 March	Day 87
	End	15 August	Day 227	31 July	Day 212
	Duration	141 Days		126 Days	
	Precipitations	866 mm (46% of total annual)		805 mm (43% of total annual)	
Short rainy season	Start	31 August	Day 243	31 August	Day 243
	End	28 November	Day 332	19 November	Day 323
	Duration	90 Days		81 Days	
	Precipitations	542 mm (29% of total annual)		486 mm (26% of total annual)	
Total	Duration	231 Days		207 hours	
	Precipitations	1408 mm (75% of total annual)		1291 mm (69% of total annual)	

Table 2.
Characteristics of rainy seasons of the southern region.

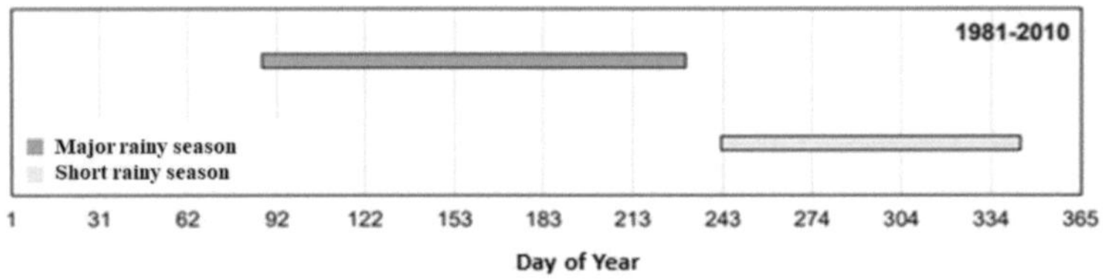

Figure 8.
Rainy season, based on initial definition of ONSET and cessation of rainy season for the southern region.

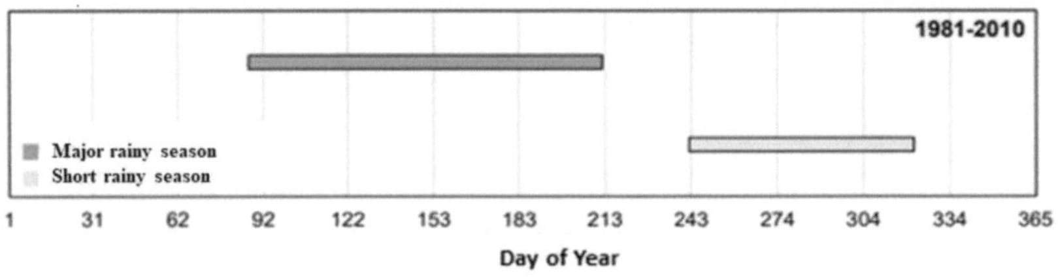

Figure 9.
Rainy seasons, based on revised definition of ONSET and cessation of rainy season for the southern region.

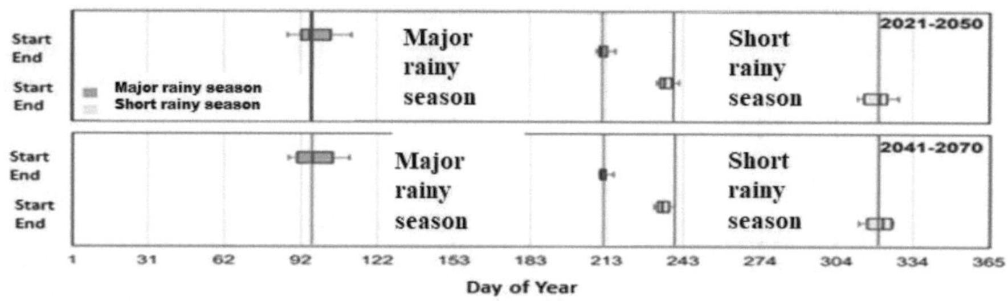

Figure 10.
Rainy seasons, based on revised parameters for the central region The vertical limits of the boxes correspond to the 10th, 50th, and 90th percentiles, while the ends of the horizontal bars indicate the minimum and maximum values for each of the climate indices associated with the rainy seasons. The vertical red lines correspond to the days of the year for the reference period, i.e. 97 (7 April), 212 (31 July), 239 (27 August), and 321 (17 November).

Using these rainy season detection parameters, the main rainy season extends from 28 March to 31 July, on average (**Table 2**). This corresponds to duration of 126 days. The short season runs from 31 August to 19 November, lasting 81 days. Note that there is a dry season between the two rainy seasons with the revised parameters.

3.3 Future climate in the centre regions

3.3.1 Temperature

The following trends can be observed:

- Average temperature: The average value for the period 1981–2010 is 25.7°C. It is expected to reach between [26.3; 26.9] °C by 2021–2050 and between [26.6; 27.7] °C by 2041–2070, representing potential increases of between [+0.6; +1.2] °C and between [+0.9; +2.0] °C, respectively.

- The average minimum daily temperature for the period 1981–2010 is 22.6°C. It is expected to rise to between [23.2; 23.9] °C by 2021–2050 and between [23.5; 24.8] °C by 2041–2070, representing potential increases of between [+0.6; +1.3] °C and between [+0.9; +2.2] °C, respectively.

- Average daily maximum temperature: The average value for the period 1981–2010 is 30.3°C. It is expected to reach between [31.0; 31.6] °C by 2021–2050 and between [31.3; 32.4] °C by 2041–2070, representing potential increases of between [+0.7; +1.3] °C and between [+1.0; +2.1] °C, respectively.

- Number of months per year with an average low daily minimum temperature (below 15°C): The average value for the period 1981–2010 is zero months/year. It would still be zero months/year by 2021–2050 and 2041–2070.

- Number of days per year with a high average daily maximum temperature (greater than or equal to 33°C): The average value for the period 1981–2010 is 51 days. This would rise to between [77; 106] days/year by 2021–2050 and between [92; 142] days/year by 2041–2070, representing potential increases of between [+26; +55] days/year and between [+41; +91] days/year, respectively.

All the scenarios and climate indices suggest warming over the next few decades. This warming would be greater for the RCP 8.5 scenario group than for the RCP 4.5 group. Also, the difference between the RCP groups would be greater for the most distant period. The most favourable temperature for growing cocoa is between 18 and 32°C. The analyses show that the number of months with an average minimum daily temperature below 15°C is zero. It is more the average maximum daily temperature that is likely to be problematic for cocoa growing. It could reach this threshold by 2021–2050 and even exceed it by 2041–2070. In addition, the number of days with an average daily maximum temperature of over 33°C could double or even triple by 2041–2070. The spatial distribution of temperatures and indices is similar.

3.3.2 Precipitation

The following trends can be observed:

- Cumulative annual precipitation: the average value for the period 1981–2010 is 1328 mm. It would be between [1317; 1491] mm by 2021–2050 and between [1315; 1533] mm by 2041–2070, representing potential changes of between [−11; +163] mm and between [−13; +205] mm, respectively.

- The number of rainy days (≥5 mm) per year; the average value for the 1981–2010 period is 89 days. It would be between [83; 99] days by 2021–2050 and between [81; 100] days by 2041–2070, representing potential changes of between [−6; +10] days and between [−8; +11] days, respectively.

- The duration of dry periods between March and November; the average value for the 1981–2010 period is 3 days. It would be between [1; 5] days by 2021–2050 and between [1; 4] days by 2041–2070, representing potential changes of between [−2; +2] days and between [−2; +1] days, respectively.

- Duration of dry spells between November and March: The average value for the 1981–2010 period is 129 days. It would be between [123; 138] days by 2021–2050 and between [123; 139] days by 2041–2070, representing potential changes of between [−6; +9] days and between [−6; +10] days, respectively.

3.3.3 Major rainy season

The following trends can be observed:

- Start of season: The season generally begins on day 97 of the year during the 1981–2010 period. It would begin between days [92; 104] of the year by 2021–2050 and between days [90; 105] of the year by 2041–2070, representing potential changes between [−5; +7] days and between [−7; +8] days. respectively.

- End of season: The season ends on day 212 of the year over the period 1981–2010. It would end between days [210; 213] of the year by 2021–2050 and between days [210; 213] of the year by 2041–2070, representing potential changes between [−2; +1] days and [−2; +1] days, respectively.

- Season length: The season lasts an average of 116 days over the 1981–2010 period. It would last between [107; 120] days by 2021–2050 and between [108; 121] days by 2041–2070, representing potential changes of between [−9; +4] days and between [−8; +5] days, respectively.

- Seasonal cumulative precipitation: Cumulative precipitation averaged 524 mm over the period 1981–2010. It would change to a value between [481; 639] mm by 2021–2050 and to a value between [483; 664] mm by 2041–2070, representing potential changes between [−43; +115] mm and between [−41; +140] mm, respectively.

3.3.4 Short rainy season

The following trends were observed:

- Start of season: The season starts on day 239 of the year over the period 1981–2010. It would begin between days [234; 239] of the year by 2021–2050 and between days [233; 238] of the year by 2041–2070, representing potential changes between [−5; 0] days and between [−6; −1] days, respectively.

- End of season: The season ends on average day 321 of the year over the period 1981–2010. It would end between days [315; 324] of the year by 2021–2050 and between days [316; 326] of the year by 2041–2070, representing potential changes between [−6; +3] days and between [−5; +5] days, respectively.

- Season length: The average season lasts 82 days over the period 1981–2010. It would last between [79; 89] days by 2021–2050 and between [79; 91] days by 2041–2070, representing potential changes of between [−3; +7] days and between [−3; +9] days, respectively.

- Seasonal cumulative precipitation: Cumulative precipitation averaged 392 mm over the period 1981–2010. It would change to between [347; 497] mm by 2021–2050 and to between [353; 504] mm by 2041–2070, representing potential changes of between [−45; +105] mm and between [−39; +112] mm, respectively.

3.3.5 Little dry season

The rainy seasons are shown in **Figure 10**. The average duration of the break between rainy seasons or little dry season is around 8 days over the 1981–2010 period, based on thresholds of 5 mm/day for 20 days. The break would last between [3; 10] days for the 2021–2050 and 2041–2070 horizons, representing potential changes between [−5; +2] days. The length of the break between rainy seasons averaged 27 days over the 1981–2010 period, based on thresholds of 10 mm/day for 10 days. The break would last between [20; 27] days by 2021–2050 and between [21; 27] days by 2041–2070, representing potential changes between [−7; 0] and between [−6; 0] days, respectively. The pause would therefore be shorter than in the present climate.

3.4 Future climate in the south region of Côte d'Ivoire

3.4.1 Temperature

The following trends were observed:

- Average temperature: The average value for the period 1981–2010 is 25.2°C. It is expected to reach between [25.8; 26.4] °C by 2021–2050 and between [26.1; 27.2] °C by 2041–2070, representing potential increases between [+0.6; +1.2] °C and between [+0.9; +2.0] °C, respectively.

- Mean minimum daily temperature: The mean value for the period 1981–2010 is 23.0°C. It is expected to reach between [23.6; 24.2] °C by 2021–2050 and between [23.9; 25.0] °C by 2041–2070, representing potential increases between [+0.6; +1.2] °C and between [+0.9; +2.0] °C, respectively.

- Mean maximum daily temperature: The mean value for the period 1981–2010 is 28.8°C. It is expected to reach between [29.4; 30.0] °C by 2021–2050 and between [29.7; 30.8] °C by 2041–2070, representing potential increases of between [+0.6; +1.2] °C and between [+0.9; +2.0] °C, respectively.

- Number of months per year with a low average daily minimum temperature (below 15°C): The average value for the period 1981–2010 is zero months/year. It would still be zero months/year by 2021–2050 and 2041–2070. Number of days per year with a high average daily maximum temperature (greater than or equal to 33°C): The average value for the period 1981–2010 is 2 days. It would rise to between [7; 17] days/year by 2021–2050 and between [12; 47] days/year by 2041–2070, representing potential increases of between [+5; +15] days/year and between [+10; +45] days/year, respectively.

All the scenarios and climate indices suggest warming over the next few decades. This warming would be greater for the RCP 8.5 scenario group than for the RCP 4.5 group. Also, the difference between the RCP groups would be greater for the most distant period. As previously mentioned, the most favourable temperature for cocoa cultivation is between 18 and 32°C. The results indicate that the number of months with an average daily minimum temperature below 15°C is zero. Concerning the upper limit of 32°C associated with cocoa cultivation, the results indicate that the average maximum daily temperature (between [29.4; 30.8]°C) would not exceed this threshold in the climatic horizons considered in this study. However, this does not mean that, in some years, the annual average would not exceed this threshold. In addition, the number of days with an average daily maximum temperature of over 33° C, which was only 2 days for the 1981–2010 period, could increase significantly. The cocoa tree's comfort limit would be exceeded more often by 2041–2070.

3.4.2 Precipitation

The following trends were observed:

- Cumulative annual precipitation: The average value for the period 1981–2010 is 1888 mm. This would rise to between [1800; 2115] mm by 2021–2050 and between [1812; 2164] mm by 2041–2070, representing potential changes of between [−88; +227] mm and between [−76; +276] mm, respectively.

- Number of rainy days (≥5 mm) per year: The average value for the 1981–2010 period is between 130 days. This would rise to between [115; 152] days by 2021–2050 and between [112; 154] days by 2041–2070, representing potential changes of between [−15; +22] days and between [−18; +24] days, respectively.

- Duration of dry periods between March and November: The average value for the 1981–2010 period is 0 days. It would still be zero months/year by 2021–2050 and 2041–2070.

- Duration of dry periods between November and March: The average value for the 1981–2010 period is 87 days. It would reach between [54; 112] days by 2021–2050 and between [55; 115] days by 2041–2070, representing potential changes of between [−33; +25] days and between [−32; +28] days, respectively.

3.4.3 Major rainy season

The following trends were observed:

- Start of season: The season starts on day 87 of the year over the period 1981–2010. It would begin between days [84; 93] of the year by 2021–2050 and between days [85; 96] of the year by 2041–2070, representing potential changes between [−3; +6] days and between [−2; +9] days, respectively.

- End of season: The season ends on day 212 of the year for the period 1981–2010. It would end between days [209; 216] of the year by 2021–2050 and between days [209; 214] of the year by 2041–2070, representing potential changes between [−3; +4] days and between [−3; +2] days, respectively.

- Season length: The season lasts an average of 125 days over the 1981–2010 period. It would last between [116; 131] days by 2021–2050 and between [115; 128] days by 2041–2070, representing potential changes of between [−9; +6] days and between [−10; +3] days, respectively.

- Seasonal cumulative precipitation: Cumulative precipitation averaged 805 mm over the period 1981–2010. It would change to a value between [768; 979] mm by 2021–2050 and to a value between [772; 1024] mm by 2041–2070, representing potential changes between [−37; +174] mm and between [−33; +219] mm, respectively.

3.4.4 Little rainy season

The following trends are observed:

- Start of season: The season begins on day 243 of the year during the 1981–2010 period. It would begin between days [237; 245] of the year by 2021–2050 and between days [235; 244] of the year by 2041–2070, representing potential changes between [−6; +2] days and between [−8; +1] days, respectively.

- End of season: The season ends on average on day 323 of the year over the period 1981–2010. It would end between days [317; 326] of the year by 2021–2050 and between days [318; 327] of the year by 2041–2070, representing potential changes between [−6; +3] days and between [−5; +4] days, respectively.

- Season length: The season lasts an average of 79 days over the 1981–2010 period. It would last between [75; 88] days by 2021–2050 and between [78; 89] days by 2041–2070, representing potential changes between [−4; +9] days and between [−1; +10] days, respectively.

- Seasonal cumulative precipitation: Cumulative precipitation averaged 486 mm over the period 1981–2010. It would change to between [384; 622] mm by 2021–2050 and to between [398; 650] mm by 2041–2070, representing potential changes of between [−102; +136] mm and between [−88; +164] mm, respectively.

3.4.5 Little dry season

The rainy seasons are shown in **Figure 10**. The average duration of the little dry season the break between rainy seasons is around 16 days over the period 1981–2010, based on thresholds of 5 mm/day for 20 days. The break would last between [6; 21] days by 2021–2050 and between [6; 20] days by 2041–2070, representing potential changes between [−10.0; +5.0] and between [−10.0; +4.0] days, respectively. The length of the break between rainy seasons averages 31 days over the 1981–2010 period, based on thresholds of 10 mm/day for 10 days. The break would last between [19; 35] days by 2021–2050 and between [21; 35] days by 2041–2070, representing potential changes between [−12; +4] and between [−10; +4] days, respectively (**Figure 11**).

4. Influence of climatic variables on cocoa tree physiology and post-harvest aspects

The aim was to establish the link between the fluctuation of climatic variables during the periods 2021 to 2050 and 2041 to 2070, on the establishment of orchards, the survival of cocoa trees after planting, the growth and development of cocoa trees, flowering, fruiting, drying and the technological quality of merchantable beans and cocoa pests and diseases. About cocoa diseases, the influence of climatic variables was

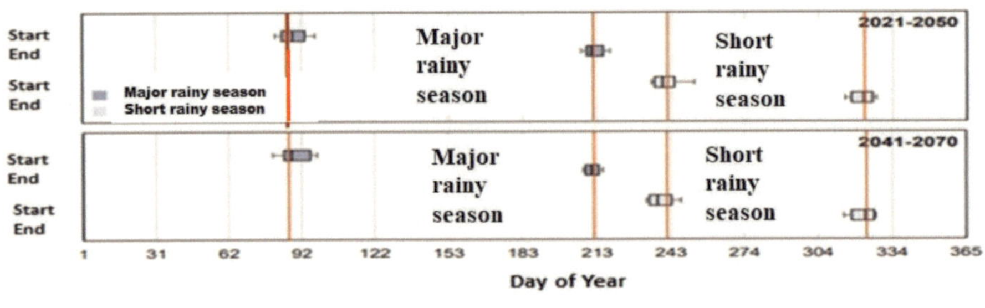

Figure 11.
Rainy seasons, based on revised parameters for the southern region. The vertical limits of the boxes correspond to the 10th, 50th and 90th percentiles, while the ends of the error bars indicate the minimum and maximum values for each of the climate indices associated with the rainy seasons. The vertical red lines correspond to the days of the year for the reference period, i.e., 87 (28 March), 212 (31 July), 243 (31 August) and 323 (19 November).

analysed on the development of brown rot and swollen shoot. Concerning pests, the analysis focused on mirids, stem borers, mealybugs, psyllids, and defoliating caterpillars. The immediate and lasting repercussions of climate change on diseases and pests are among the major concerns of scientists and policy-makers around the world. Rainfall and temperature are the most important factors in the development of diseases and pests.

Tables 3–10 summarise the results of the climate scenarios and their influence on cocoa production in the Central and Southern zones of Côte d'Ivoire.

5. Identification of climate smart practices for adaptation to climate variability in cocoa production

Generally, the adoption of climate-smart practices by cocoa farmers remains limited for a variety of reasons. Surveys carried out in Abengourou, Gagnoa, Vavoua, and Soubré to collect and analyse adaptation practices in cocoa farming enabled a range of agroecological practices covering all stages of cultivation to be inventoried (**Table 11**). The most promising practices for coping with the effects of climate change in cocoa farming were identified using multiple criteria (economic, social, and gender equality, environmental, and efficiency). This multi-criteria analysis shows that the most promising practices relate to tree planting, for shade purposes and other benefits provided by trees and agroforestry [27, 28]. These practices deserve to be included in the training curricula of professionals who work with cocoa farmers, just as these practices deserve to be widely disseminated and promoted among cocoa farmers. Cocoa cooperatives also play a role in disseminating these messages and in meeting the needs of their members, both men and women. These organizations benefit from taking climate change into account in the products and services they offer so that they can make informed and strategic decisions.

6. Conclusions

The analysis of the climate scenarios for the next 50 years for the central and southern zones of Côte d'Ivoire concerning cocoa farming provides an understanding of the projected impact of future climate on cocoa trees and the surrounding environment. In general, the various climate scenarios showed that there would be an increase in the average maximum temperature (up to 1.3°C higher by 2021–2050 and 2.2°C higher by 2041–2070) and an increase in the number of hot days (≥33°C) in both zones (Centre and South), although these changes would be more marked in the Centre zone. No consensus has been reached between the climate model projections for rainfall, except that the short rainy season could start earlier than in the present climate. These predictions could have impacts on cocoa cultivation. With respect to physiology, rainfall forecasts are favourable to the establishment, growth, flowering, and fruiting of cocoa trees. On the other hand, an increase in the number of hot days would be harmful to the cocoa tree. Concerning cocoa diseases (CSSVD and brown rot), increases in temperature and the number of hot days would help slow their progress in the orchards. About the main pests, increases in temperature and rainfall would not alter their dynamics, but only the delay in the start of the main rainy season could reduce the outbreak of defoliating caterpillars. For soil on

	Central zone: period from 2021 to 2050				**Cocoa physiology and post-harvest aspects**						
Parameters	**Reference period values 1981–2010**	**Normal values for cacao**	**Projections 2021–2050 (RCP 4.5 et RCP 8.5)**	**Changes from 2021 to 2050/ reference period**	**Establishment in the farm**	**Growth and development**	**Flowering**	**Fructification**	**Production**	**Quality (size) of beans**	**Drying**
Temperature											
Average temperature (°C)	25.7 °C	25–26 °C	[26.3; 26.9]	[+0.6; +1.2]	Favourable	Favourable	Favourable	Favourable	Favourable	Favourable	Favourable
Minimum temperature (°C)	22.6 °C	18 °C	[23.2; 23.9]	[+0.6; +1.3]	Favourable	Favourable	Favourable	Favourable	Favourable	Favourable	Favourable
Maximum temperature (°C)	30.3 °C	32 °C	[31.0; 31.6]	[+0.7; +1.3]	Favourable	Favourable	Favourable	Favourable	Favourable	Little favourable	Favourable
Number of warm days (≥ 33 °C)	51 Day/ year	51 Days/ Year	[77; 106]	[+26; +55]	Unfavourable	Unfavourable	Unfavourable	Unfavourable	Unfavourable	Little favourable	Favourable
Precipitation											
Rainfall (total annual, (mm)	1328 mm	[1200; 2000]	[1317; 1491]	[−11; +163]	Favourable	Favourable	Favourable	Favourable	Favourable	Favourable	Favourable
Number of rainy days/year (≥5 mm/ day)	89 Days	89 Days	[83; 99]	[−6; +10] days	Favourable	Favourable	Favourable	Favourable	Favourable	Favourable	Favourable
Number of dry days (<1 mm/day over 14 days)	3 Days	3 Days	[1; 5] days	[−2; +2] days	Favourable	Favourable	Favourable	Favourable	Favourable	Favourable	Favourable
Start of the major rainy season	07-Apr	15-May	[02 Apr; 14 Apr]	[−5; +7] days	Favourable	Favourable	Favourable	Favourable	Favourable	Favourable	Favourable
End of major rainy season (<10 mm over 10 days)	31-Jul	15-Jul	[29 July; 1er Aug]	[−2; +1] days	Favourable	Favourable	Favourable	Favourable	Favourable	Favourable	Favourable

DOI: http://dx.doi.org/10.5772/intechopen.112643

	Central zone: period from 2021 to 2050				Cocoa physiology and post-harvest aspects						
Parameters	**Reference period values 1981–2010**	**Normal values for cacao**	**Projections 2021–2050 (RCP 4.5 et RCP 8.5)**	**Changes from 2021 to 2050/ reference period**	**Establishment in the farm**	**Growth and development**	**Flowering**	**Fructification**	**Production**	**Quality (size) of beans**	**Drying**
Duration of major rainy season	116 Days	122 Days (700 mm)	[107 Days; 120 days]	[−9; +4] days	Favourable	Favourable	Favourable	Favourable	Favourable	Favourable	Favourable
Start of short rainy season	26-Aug	15-Sept	[22 to 27 Aug]	[−4; +1] days	Favourable	Favourable	Favourable	Favourable	Favourable	Favourable	Favourable
End of short rainy season (<10 mm over 10 days)	17-Nov	11-Nov	[11–20 Nov]	[−6; +3] days	Favourable	Favourable	Favourable	Favourable	Favourable	Favourable	Favourable
Duration of short rainy season	82 Days	65 Days (70 mm)	[79 to 89 Days]	[−3; +7] days	Favourable	Favourable	Favourable	Favourable	Favourable	Favourable	Favourable

Table 3.
Interpretation of the influence of climatic variables for the period 2021–2050 on cocoa physiology and post-harvest aspects in the central zone.

Central zone: period from 2041 to 2050					Cocoa physiology and post-harvest aspects							
Parameters	**Reference period values 1981–2010**	**Normal values for cacao**	**Projections 2041–2070 (RCP 4.5 et RCP 8.5)**	**Changes from 2021 to 2050/ reference period**	**Establishment in the farm**	**Growth and development**	**Flowering**	**Fructification**	**Production**	**Quality (size) of beans**	**Drying**	**Storage**
Temperature												
Average temperature (°C)	25.7 °C	25–26 °C	[26.6; 27.7]	[+0.9; +2.0]	Favourable	Favourable	Favourable	Favourable	Favourable	Favourable	Favourable	Favourable
Minimum temperature (°C)	22.6 °C	18 °C	[23.5 ; 24.8]	[+0.9 ; +2.2]	Favourable	Favourable	Favourable	Favourable	Favourable	Favourable	Favourable	Favourable
Maximum temperature (°C)	30.3 °C	32 °C	[31.3 ; 32.4]	[+1.0 ; +2.1]	Favourable	Favourable	Favourable	Favourable	Favourable	Favourable	Favourable	Favourable
Number of warm days (≥ 33 °C)	51 days/ year	51 days/ year	[92 ; 142]	[+41 ; +91]	Favourable	unfavourable	unfavourable	unfavourable	unfavourable	unfavourable	unfavourable	unfavourable
Precipitation												
Rainfall (total annual, (mm)	1328 mm	[1200 ; 2000]	[1315 ; 1533]	[−13 ; +205]	Favourable	Favorable	Favorable	Favorable	Favorable	Favorable	Favorable	Favorable
Number of rainy days/year (≥5 mm/ day)	89 Days	89 Days	[81 ; 100]	[−8 ; +11]	Favourable	Favorable	Favorable	Favorable	Favorable	Favorable	Favorable	Favorable
Number of dry days (<1 mm/day over 14 days)	3 Days	3 Days	[1 ; 4]	[−2 ; +1]	Favourable	Favourable	Favourable	Favourable	Favourable	Favourable	Favourable	Favourable
Start of the major rainy season	07-Apr	15-Mar	[31 Mar ; 15 April]	[−7 ; +8]	Favourable	Favourable	Favourable	Favourable	Favourable	Favourable	Favourable	Favourable
End of major rainy season (<10 mm over 10 days)	31-Jul	15-Jul	[29 Jul; 1er Aug]	[−2 ; +1]	Favourable	Favourable	Favourable	Favourable	Favourable	Favourable	Favourable	Favourable

	Central zone: period from 2041 to 2050				Cocoa physiology and post-harvest aspects							
Parameters	**Reference period values 1981–2010**	**Normal values for cacao**	**Projections 2041–2070 (RCP 4.5 et RCP 8.5)**	**Changes from 2021 to 2050/ reference period**	**Establishment in the farm**	**Growth and development**	**Flowering**	**Fructification**	**Production**	**Quality (size) of beans**	**Drying**	**Storage**
Duration of major rainy season	116 Days	122 days (700 mm)	[108; 121] Days	[−8 ; +1] Days	Favourable	Favourable	Favourable	Favourable	Favourable	Favourable	Favourable	Favourable
Start of short rainy season	26-Aug	15 Sept.	[21–26 Aug]	[−5; +0]	Favourable	Favourable	Favourable	Favourable	Favourable	Favourable	Favourable	Favourable
End of short rainy season (<10 mm over 10 days)	17-Nov	11-Nov	[12–22 Nov]	[−5; +5]	Favourable	Favourable	Favourable	Favourable	Favourable	Favourable	Favourable	Favourable
Duration of short rainy season	82 Days	65 Days (70 mm mm)	[79; 83] Days	[−3 ; +1] Days	Favorable	Favorable	Favorable	Favorable	Favorable	Favorable	Favorable	Favorable

Table 4.
Interpretation of the influence of climatic variables for the period 2041–2070 on cocoa physiology and post-harvest aspects in the central zone.

	Central zone: period from 2021 to 2050				Diseases and pests of the cacao						
Parameters	**Reference period 1981–2010**	**Normal values for cacao**	**Projections 2021–2050 (RCP 4.5 and RCP 8.5)**	**Change from 2021 to 2050/ reference period**	**CSSVD**	**Black pod**	**Mirids**	**Stem borers**	**Mealybugs**	**Psylles**	**Defoliating caterpillars**
Temperature											
Average temperature (°C)	25.7 °C	25–26 °C	[26.3; 26.9]	25.7 °C	Favourable	Favourable	Favourable	Favourable	Favourable	Favourable	Favourable
Minimum temperature (°C)	22.6 °C	18 °C	[23.2; 23.9]	22.6 °C	Favourable	Favourable	Favourable	Favourable	Favourable	Favourable	Favourable
Maximum temperature (°C)	30.3 °C	32 °C	[31.0; 31.6]	30.3 °C	Favourable	Little favourable	Favourable	Favourable	Favourable	Favourable	Favourable
Number of warm days (≥ 33 °C)	51 Days/ year	51 Days/ year	[77 ; 106]	51 days/year	Unfavourable	Unfavourable	Little favourable	Little favourable	Little favourable	Little favourable	Little favourable
Precipitation											
Rainfall (total annual, (mm)	1328 mm	[1200 ; 2000]	[1317 ; 1491]	1328 mm	Favourable	Favourable	Favourable	Favourable	Favourable	Favourable	Favourable
Number of rainy days/year (≥5 mm/day)	89 Days	89 Days	[83 ; 99] Days	89 Days	Favourable	Favourable	Favourable	Favourable	Favourable	Favourable	Favourable
Number of dry days (<1 mm/ day over 14 days)	3 Days	3 Days	[1; 5] Days	[−2; +2]	Favourable	Little favourable	Favourable	Favourable	Favourable	Favourable	Favourable
Start of the major rainy season	07-Avr	15-Mar	[02–14 April]	[−5; +7] Days	Favourable	Favourable	Favourable	Favourable	Favourable	Favourable	Little favourable
End of major rainy season (<10 mm over 10 days)	31-Jul	15-Jul	[29 Jul; 1er Aug]	[−2; +3] Days	Favourable	Favourable	Favourable	Favourable	Favourable	Favourable	Favourable

	Central zone: period from 2021 to 2050				Diseases and pests of the cacao						
Parameters	**Reference period 1981–2010**	**Normal values for cacao**	**Projections 2021–2050 (RCP 4.5 and RCP 8.5)**	**Change from 2021 to 2050/ reference period**	**CSSVD**	**Black pod**	**Mirids**	**Stem borers**	**Mealybugs**	**Psylles**	**Defoliating caterpillars**
Duration of major rainy season	116 Days	122 Days	[107 – 120 Days]	[−9; +3] Days	Favourable	Favourable	Favourable	Favourable	Favourable	Favourable	Favourable
Start of short rainy season	26-Aug	15-Sept	[22 to 27 Aug]	[−4; +1] Days	Favourable	Favourable	Favourable	Favourable	Favourable	Favourable	Favourable
End of short rainy season (<10 mm over 10 days)	17-Nov	11-Nov	[11–20 Nov]	[−6; +3] Days	Favourable	Favourable	Favourable	Favourable	Favourable	Favourable	Favourable
Duration of short rainy season	82 Days	65 Days (70 mm)	[79 to 89 Days]	[−3; +7]Days	Favourable	Favourable	Favourable	Favourable	Favourable	Favourable	Favourable

Table 5.
Interpretation of the influence of climatic variables for the period 2021–2050 on cacao diseases and pests in the central zone.

Central zone: period from 2041 to 2070					Diseases and pests of the cacao						
Parameters	**Values for the reference period 1981–2010**	**Normal values for cacao**	**Projections 2041 –2070 (RCP 4.5 and RCP 8.5)**	**Change from 2041 to 2070/ reference period**	**CSSVD**	**Black pod**	**Mirids**	**Stem borers**	**Mealybugs**	**Psylles**	**Defoliating caterpillars**
Temperature											
Average temperature (°C)	25.7 °C	25–26 °C	[26.6; 27.7]	[+0.9; +2.0]	Favourable	Favourable	Favourable	Favourable	Favourable	Favourable	Favourable
Minimum temperature (°C)	22.6 °C	18 °C	[23.5 ; 24.8]	[+0.9 ; +2.2]	Favourable	Favourable	Favourable	Favourable	Favourable	Favourable	Favourable
Maximum temperature (°C)	30.3 °C	32 °C	[31.3 ; 32.4]	[+1.0 ; +2.1]	Favourable	Little favourable	Favourable	Favourable	Favourable	Favourable	Favourable
Number of warm days (≥ 33 °C)	51 Days/year	51 Days/ year	[92 ; 142]	[+41 ; +91]	unfavourable	unfavourable	Little favourable	Little favourable	Little favourable	Little favourable	Little favourable
Precipitation											
Rainfall (total annual, (mm)	1328 mm	[1200 ; 2000]	[1315 ; 1533]	[−13 ; +205]	Favourable	Favourable	Favourable	Favourable	Favourable	Favourable	Favourable
Number of rainy days/year (≥5 mm/day)	89 Days	89 Days	[81 ; 100]	[−8; +11]	Favourable	Favourable	Favourable	Favourable	Favourable	Favourable	Favourable
Number of dry days (<1 mm/day over 14 days)	3 Days	3 Days	[1 ; 4]	[−2 ; +1]	Favourable	Little favourable	Favourable	Favourable	Favourable	Favourable	Favourable
Start of the major rainy season	06-Apr	15-Mar	(30 Mar ; 14 Apr)	[−7 ; +8]	Favourable	Favourable	Favourable	Favourable	Favourable	Favourable	Little favourable
End of major rainy season (<10 mm over 10 days)	30-Jul	15-Jul	(28–31 Jul)	[−2 ; +1]	Favourable	Favourable	Favourable	Favourable	Favourable	Favourable	Favourable

Central zone: period from 2041 to 2070					Diseases and pests of the cacao						
Parameters	**Values for the reference period 1981–2010**	**Normal values for cacao**	**Projections 2041 –2070 (RCP 4.5 and RCP 8.5)**	**Change from 2041 to 2070/ reference period**	**CSSVD**	**Black pod**	**Mirids**	**Stem borers**	**Mealybugs**	**Psylles**	**Defoliating caterpillars**
Duration of major rainy season	116 Days	122 Days (700 mm)	[108; 121] Days	[−8 ; +1]	Favourable	Favourable	Favourable	Favourable	Favourable	Favourable	Favourable
Start of short rainy season	26-Agus	15-Sept	[21 26 Aug]	[−5; +0]	Favourable	Favourable	Favourable	Favourable	Favourable	Favourable	Favourable
End of short rainy season (<10 mm over 10 days)	17-Nov	11-Nov	[12 ; 22 Nov]	[−5; +5]	Favourable	Favourable	Favourable	Favourable	Favourable	Favourable	Favourable
Duration of short rainy season	82 Days	65 Days (70 mm)	[79; 83] Days	[−3 ; +1]	Favourable	Favourable	Favourable	Favourable	Favourable	Favourable	Favourable

Table 6.
Interpretation of the influence of climatic variables for the period 2041–2070 on cocoa diseases and pests in the central zone.

	Southern zone: period 2021 à 2050				Cocoa physiology and post-harvest aspects							
Parameters	**Reference period values 1981–2010**	**Normal values for cacao**	**Projections 2021–2050 (RCP 4.5 and RCP 8.5)**	**Changes from 2021 to 2050/ reference period**	**Establishment in the farm**	**Growth and development**	**Flowering**	**Fructification**	**Production**	**Quality (size) of beans**	**Drying**	**Storage**
Temperature												
Average temperature (°C)	25.2 °C	25–26 °C	[25.8; 26.4]	[+0.6; +1.2]	Favourable	Favourable	Favourable	Favourable	Favourable	Favourable	Favourable	Favourable
Minimum temperature (°C)	23.0 °C	18 °C	[23.6 ; 24.2]	[+0.6 ; +1.2]	Favourable	Favourable	Favourable	Favourable	Favourable	Favourable	Favourable	Favourable
Maximum temperature (°C)	28.8 °C	32 °C	[29.4 ; 30.0]	[+0.6 ; +1.2]	Favourable	Favourable	Favourable	Favourable	Favourable	Favourable	Favourable	Favourable
Number of warm days (≥ 33 °C)	2 Days/ year	2 Days/ year	[7 ; 17]	[+5 ; +15]	unfavourable	unfavourable	unfavourable	unfavourable	unfavourable	unfavourable	unfavourable	unfavourable
Precipitation												
Rainfall (total annual, (mm)	1888 mm	[1200 ; 2000]	[1800 ; 2115]	[−88 ; +227]	Favourable	Favourable	Favourable	Favourable	Favourable	Favourable	Favourable	Favourable
Number of rainy days/year (≥5 mm/day)	130	130 jours	[115 ; 152]	[−15 ; +22]	Favourable	Favourable	Favourable	Favourable	Favourable	Favourable	Favourable	Favourable
Number of dry days (<1 mm/day over 14 days)	0 day	0 day	[0 ; 0]	[0 ; 0]	Favourable	Favourable	Favourable	Favourable	Favourable	Favourable	Favourable	Favourable
Start of the major rainy season	27-Mar	15-Mar	[25 Mar ; 03 Apr]	[−2 ; +7] Days	Favourable	Favourable	Favourable	Favourable	Favourable	Favourable	Favourable	Favourable
End of major rainy season (<10 mm over 10 days)	30-Jul	15-Jul	[28 Jul; 4 Aug]	[−2 ; +5] Days	Favourable	Favourable	Favourable	Favourable	Favourable	Favourable	Favourable	Favourable
Duration of major rainy season	125 Days	122 Days (700 mm)	[116; 131] Days	[−9 ; +6] Days	Favourable	Favourable	Favourable	Favourable	Favourable	Favourable	Favourable	Favourable

	Southern zone: period 2021 à 2050				Cocoa physiology and post-harvest aspects							
Parameters	**Reference period values 1981–2010**	**Normal values for cacao**	**Projections 2021–2050 (RCP 4.5 and RCP 8.5)**	**Changes from 2021 to 2050/ reference period**	**Establishment in the farm**	**Growth and development**	**Flowering**	**Fructification**	**Production**	**Quality (size) of beans**	**Drying**	**Storage**
Start of short rainy season	31-Aug	15-Sept	[23 Aug; 01 Sept]	[−8; +1] Days	Favourable	Favourable	Favourable	Favourable	Favourable	Favourable	Favourable	Favourable
End of short rainy season (<10 mm over 10 days)	19-Nov	11-Nov	[14 to 23 Nov]	[−5; +4] Days	Favourable	Favourable	Favourable	Favourable	Favourable	Favourable	Favourable	Favourable
Duration of short rainy season	79 Days	65 Days (70 mm)	[75; 88] Days	[−4; +9] Days	Favourable	Favourable	Favourable	Favourable	Favourable	Favourable	Favourable	Favourable

Table 7.
Interpretation of the influence of climatic variables for the period 2021–2050 on cocoa physiology and post-harvest aspects in the southern zone.

	South zone: period 2041 to 2070				Cacao physiology and post-harvest aspects							
Parameters	**Reference period values 1981–2010**	**Normal values for cacao**	**Projections 2041–2070 (RCP 4.5 et RCP 8.5)**	**Changes from 2021 to 2050/ reference period**	**Establishment in the farm**	**Growth and development**	**Flowering**	**Fructification**	**Production**	**Quality (size) of beans**	**Drying**	**Storage**
Temperature												
Average temperature (°C)	25.2 °C	25–26 °C	[26.1; 27.2]	[+0.9; +2.0]	Favourable	Favourable	Favourable	Favourable	Favourable	Favourable	Favourable	Favourable
Minimum temperature (°C)	23.0 °C	18 °C	[23.9 ; 25.0]	[+0.9 ; +2.0]	Favourable	Favourable	Favourable	Favourable	Favourable	Favourable	Favourable	Favourable
Maximum temperature (°C)	28.8 °C	32 °C	[29.7 ; 30.8]	[+0.9 ; +2.0]	Favourable	Favourable	Favourable	Favourable	Favourable	Favourable	Favourable	Favourable
Number of warm days (≥ 33 °C)	2 Days/ year	2 Days/ year	[12 ; 47]	[+17 ; +45]	Unfavourable	Unfavourable	Unfavourable	Unfavourable	Unfavourable	Unfavourable	Unfavourable	Unfavourable
Precipitation												
Rainfall (total annual, (mm)	1888 mm	[1200; 2000]	[1812; 2164]	[−76; +276]	Favourable	Favourable	Favourable	Favourable	Favourable	Favourable	Favourable	Favourable
Number of rainy days/year (≥5 mm/ day)	130	130 Days	[112; 154]	[−18; +24]	Favourable	Favourable	Favourable	Favourable	Favourable	Favourable	Favourable	Favourable
Number of dry days (<1 mm/day over 14 days)	0 Dry days	0 dry days	[0 ; 0]	[0 ; 0]	Favourable	Favourable	Favourable	Favourable	Favourable	Favourable	Favourable	Favourable
Start of the major rainy season	27-Mar	15-Mar	[25 Mar ; 5 Apr]	[−2 ; +9]	Favourable	Favourable	Favourable	Favourable	Favourable	Favourable	Favourable	Favourable
End of major rainy season (<10 mm over 10 days)	30-Juil	15-Jul	[27 Jul; 1 Aug]	[−3 ; +2]	Favourable	Favourable	Favourable	Favourable	Favourable	Favourable	Favourable	Favourable

DOI: http://dx.doi.org/10.5772/intechopen.112643

South zone: period 2041 to 2070					Cacao physiology and post-harvest aspects							
Parameters	**Reference period values 1981–2010**	**Normal values for cacao**	**Projections 2041–2070 (RCP 4.5 et RCP 8.5)**	**Changes from 2021 to 2050/ reference period**	**Establishment in the farm**	**Growth and development**	**Flowering**	**Fructification**	**Production**	**Quality (size) of beans**	**Drying**	**Storage**
Duration of major rainy season	125 Days	122 Days (700 mm)	[115; 128] Days	[−10 ; +3] Days	Favourable	Favourable	Favourable	Favourable	Favourable	Favourable	Favourable	Favourable
Start of short rainy season	31-Aug	15-Sept	[22–31 Aug]	[−9; +0] Days								
End of short rainy season (<10 mm over 10 days)	19-Nov	11-Nov	[14–23 Nov]	[−5; +4] Days	Favourable	Favourable	Favourable	Favourable	Favourable	Favourable	Favourable	Favourable
Duration of short rainy season	79 Days	65 Days (70 mm)	[78; 89] Days	[−1; +10] Days	Favourable	Favourable	Favourable	Favourable	Favourable	Favourable	Favourable	Favourable

Table 8.
Interpretation of the influence of climatic variables for the period 2041–2070 on cacao physiology and post-harvest aspects in the southern zone.

South zone: period 2021 to 2050					Cacao diseases and pests						
Parameters	**Reference period 1981–2010**	**Normal values for cacao**	**Projections 2021–2050 (RCP 4.5 and RCP 8.5)**	**Change from 2021 to 2050/ reference period**	**CSSVD**	**Black pod**	**Mirids**	**Stem borers**	**Mealybugs**	**Psylles**	**Defoliating caterpillars**
Temperature											
Average temperature (°C)	25.2 °C	25–26 °C	[25.8; 26.4]	[+0.6; +1.2]	Favourable	Favourable	Favourable	Favourable	Favourable	Favourable	Favourable
Minimum temperature (°C)	23.0 °C	18 °C	[23.6 ; 24.2]	[+0.6 ; +1.2]	Favourable	Favourable	Favourable	Favourable	Favourable	Favourable	Favourable
Maximum temperature (°C)	28.8 °C	32 °C	[29.4 ; 30.0]	[+0.6 ; +1.2]	Favourable	Favourable	Favourable	Favourable	Favourable	Favourable	Favourable
Number of warm days (≥ 33 °C)	2 Days/ year	2 Days/ year	[7 ; 17]	[+5 ; +15]	Favourable	Favourable	Favourable	Favourable	Favourable	Favourable	Favourable
Precipitation											
Rainfall (total annual, (mm)	1888 mm	[1200 ; 2000]	[1800 ; 2115]	[−88 ; +227]	Favourable	Favourable	Favourable	Favourable	Favourable	Favourable	Favourable
Number of rainy days/year (≥5 mm/day)	130	130 Days	[115 ; 152]	[−15 ; +22]	Favourable	Favourable	Favourable	Favourable	Favourable	Favourable	Favourable
Number of dry days (<1 mm/day over 14 days)	0 Dry day	0 dry day	[0 ; 0]	[0 ; 0]	Favourable	Favourable	Favourable	Favourable	Favourable	Favourable	Favourable
Start of the major rainy season	27-Mar	15 Mar	[25 Mar; 03 Apr]	[−2 ; +7 Days	Favourable	Favourable	Favourable	Favourable	Favourable	Favourable	Favourable
End of major rainy season (<10 mm over 10 days)	30-Jul	15-Jul	[28 Jul; 4 Aug]	[−2 ; +5] Days	Favourable	Favourable	Favourable	Favourable	Favourable	Favourable	Favourable

South zone: period 2021 to 2050					Cacao diseases and pests						
Parameters	**Reference period 1981–2010**	**Normal values for cacao**	**Projections 2021–2050 (RCP 4.5 and RCP 8.5)**	**Change from 2021 to 2050/ reference period**	**CSSVD**	**Black pod**	**Mirids**	**Stem borers**	**Mealybugs**	**Psylles**	**Defoliating caterpillars**
Duration of major rainy season	125 Days	122 Days (700 mm)	[116; 131] Days	[−9 ; +6] Days	Favourable	Favourable	Favourable	Favourable	Favourable	Favourable	Favourable
Start of short rainy season	31-Aug	15-Sept	[23 Aug; 01 Sept]	[−8; +1] Days	Favourable	Favourable	Favourable	Favourable	Favourable	Favourable	Favourable
End of short rainy season (<10 mm over 10 days)	19-Nov	11-Nov	[14–23 Nov]	[−5; +4] Days	Favourable	Favourable	Favourable	Favourable	Favourable	Favourable	Favourable
Duration of short rainy season	79 Days	65 Days (70 mm)	[75; 88] Days	[−4 +9] Days	Favourable	Favourable	Favourable	Favourable	Favourable	Favourable	Favourable

Table 9.
Interpretation of the influence of climatic variables for the period 2021–2050 on cocoa diseases and pests in the southern zone.

South zone: period 2041 to 2070					Cacao diseases and pests						
Parameters	**Reference period 1981–2010**	**Normal values for cacao**	**Projections 2041–2070 (RCP 4.5 and RCP 8.5)**	**Change from 2041 to 2070/ reference period**	**CSSVD**	**Black pod**	**Mirids**	**Stem borers**	**Mealybugs**	**Psylles**	**Defoliating caterpillars**
Temperature											
Average temperature (°C)	25.2 °C	25–26 °C	[26.1; 27.2]	[+0.9; +2.0]	Favourable	Favourable	Favourable	Favourable	Favourable	Favourable	Favourable
Minimum temperature (°C)	23.0 °C	18 °C	[23.9 ; 25.0]	[+0.9 ; +2.0]	Favourable	Favourable	Favourable	Favourable	Favourable	Favourable	Favourable
Maximum temperature (°C)	28.8 °C	32 °C	[29.7 ; 30.8]	[+0.9 ; +2.0]	Favourable	Favourable	Favourable	Favourable	Favourable	Favourable	Favourable
Number of warm days (≥ 33 °C)	2 Days/ year	2 Dry/ year	[12 ; 47]	[+17 ; +45]	Unfavourable	Unfavourable	Unfavourable	Unfavourable	Unfavourable	Favourable	Favourable
Precipitation											
Rainfall (total annual, (mm)	1888 mm	[1200 ; 2000]	[1812 ; 2164]	[−76 ; +276]	Favourable	Favourable	Favourable	Favourable	Favourable	Favourable	Favourable
Number of rainy days/year (≥5 mm/day)	130 Days	130 days	[112 ; 154]	[−18 ; +24]	Favourable	Favourable	Favourable	Favourable	Favourable	Favourable	Favourable
Number of dry days (<1 mm/day over 14 days)	0 Dry day	0 Dry day	[0 ; 0]	[0 ; 0]	Favourable	Favourable	Favourable	Favourable	Favourable	Favourable	Favourable
Start of the major rainy season	27-Mar	15-Mar	[25 Mar ; 5 Apr]	[−2 ; +9]	Favourable	Favourable	Favourable	Favourable	Favourable	Favourable	Little favourable

South zone: period 2041 to 2070					Cacao diseases and pests						
Parameters	**Reference period 1981–2010**	**Normal values for cacao**	**Projections 2041–2070 (RCP 4.5 and RCP 8.5)**	**Change from 2041 to 2070/ reference period**	**CSSVD**	**Black pod**	**Mirids**	**Stem borers**	**Mealybugs**	**Psylles**	**Defoliating caterpillars**
End of major rainy season (<10 mm over 10 days)	30-Jul	15-Jul	[27 Jul; 1 Aug]	[−3 ; +2]	Favourable	Favourable	Favourable	Favourable	Favourable	Favourable	Favourable
Duration of major rainy season	125 Days	122 days (700 mm)	[115; 128] Days	[−10 ; +3] Days	Favourable	Favourable	Favourable	Favourable	Favourable	Favourable	Favourable
Start of short rainy season	31-Aug	15-Sept	[22–31 Aug]	[−9; +0] Days	Favourable	Favourable	Favourable	Favourable	Favourable	Favourable	Favourable
End of short rainy season (<10 mm over 10 days)	19-Nov	11-Nov	[14–23 Nov]	[−5; +4] Days	Favourable	Favourable	Favourable	Favourable	Favourable	Favourable	Favourable
Duration of short rainy season	79 Days	65 Days (70 mm)	[78; 89] Day	[−1; +10] Days	Favourable	Favourable	Favourable	Favourable	Favourable	Favourable	Favourable

Table 10.
Interpretation of the influence of climatic variables for the period 2041–2070 on cocoa diseases and pests in the southern zone.

Illustrations	**Agroecological practices identified**
Temporary shade plants	
	• Plantation of banana trees for temporary shade • Planting tree legumes (*Cajanus cajan*) for temporary shade and human consumption
Permanent shade trees	
	• Plantation of compatible fruit and forest species for shade purposes • Plantation of perennial legumes (e.g. *Acacia mangium* and Albizia)
Crop system management	
	• Association of cocoa with food crops (e.g. yam, tomato, okra ...) • Association of cocoa with edible cover crops (e.g. groundnuts, beans, soya, sweet potatoes)

Illustrations	**Agroecological practices identified**
Soil fertility management	
	• Production and use of compost made from cocoa residues • Mulching around cacao
Pruning cocoa trees	
	• Flowering pruning once a year in line with the start of the rainy season (after the first heavy rain) and cleaning of pruning equipment • Training pruning in the rainy season and cleaning of pruning equipment
Disease and pest management	
	• Use of vegetative barriers against swollen shoot and strong winds • Regular inspection of orchards for early detection of symptoms of disease and pests • Use of mechanical methods (pruning, plugging holes with stems, manual removal of insects, leaves, attacked fruit) to prevent the appearance of and/or control pests • Use of grass swaths to prevent attacks by harmful insects • Use of natural/biological repellents (neem, ginger, predatory insects) to prevent the appearance of and/or control pests

Illustrations	Agroecological practices identified
Post-harvest activities	
	• Storage of cocoa beans protected from bad weather

Table 11.
Identification of agroecological practices to adapt to climate change effect in cocoa farming.

which crop is grown, (cocoa trees, and associated crops) would also be affected by climate change, with an increase in the number of hot days (≥33°C). This would result in high evapotranspiration, low soil water retention, and rapid degradation of organic matter, leading to a decline in soil fertility. In addition, climate change could lead to lower yields and higher levels of poverty in rural areas. Given the impact that climate change would have on cocoa farming over the periods 2021–2050 and 2041–2070, strategies are proposed to improve the resilience of producers within cocoa cooperatives in Côte d'Ivoire. A set of agro-ecological practices have been identified that could counteract the adverse effects of climate change.

Acknowledgements

This work was done in the framework of the Adaptcoop project in Côte d'Ivoire. The authors would like to thank the IDRC (International Development Research Centre of Canada) for funding the project. We also thank SOCODEVI, and OURANOS from Canada, and WASCAL/CEA-CCBAD (African Centre of Excellence on Climate Change, Biodiversity and Sustainable Agriculture) and CNRA (Centre National of Research on Agriculture), from Côte d'Ivoire.

Conflict of interest

The authors declare no conflict of interest.

Author details

Antoine Alban Kacou M'bo[1*], Mamadou Cherif[1], Kouakou Kouadio[1,2], Mahyao Germain Adolphe[3], Adama Bamba[1,4], Evelyne N'Datchoh Toure[1,4], Alla Kouadio Okou[5], Renée Brunelle[6], Yanick Rouseau[7] and Daouda Koné[1]

1 African Center of Excellence on Climate Change, Biodiversity and Sustainable Agriculture (CEA-CCBAD/WASCAL), Université Félix Houphouet-Boigny (UFHB), Côte d'Ivoire

2 Geophysical Station of Lamto (GSL), N'Douci, Côte d'Ivoire

3 Centre National de Recherche Agronomique, Gagnoa Research Station, Côte d'Ivoire

4 Laboratoire des Sciences de la Matière, de l'environnement et de l'Energie Solaire (LASMES), UFR-SSMT of Felix Houphouët Boigny University, Côte d'Ivoire

5 Agence National d'Appui au Developpement Rurale (ANADER), Côte d'Ivoire

6 SOCODEVI, International Cooperation Organization, Canada

7 OURANOS, Consortium on Regional Climatology and Adaptation to Climate Change Montréal, Québec, Canada

*Address all correspondence to: mbo.antoine@ufhb.edu.ci

References

[1] ICCO Quarterly Bulletin of Cocoa Statistics. XLVII. No. 2. 2020/21. Available from: https://www.icco.org/icco-documentation/quarterly-bulletin-of-cocoa-statistics/

[2] UTZ Certified et Solidaridad. The Role of Certification and Producer Support in Promoting Gender Equality in Cocoa Production. 2009. Récupéré sur: https://www.utzcertified.org/attachments/article/92/CocoaGenderUTZSolidaridad_2009.pdf

[3] Assiri AA. Etude de la régénération cacaoyère en côte d'ivoire : impact des techniques de réhabilitation et de replantation sur le développement et la productivité des vergers de cacaoyers (*Theobroma cacao* L.) en relation avec l'état du sol. In: Thèse de doctorat, UFR STRM option Agro-pédologie, Université de Cocody. Abidjan: Côte d'Ivoire; 2010. p. 170

[4] Koffié K, Kébé BI, Kouassi N, Anno AP, Ake S, Muller E. Impact de la maladie virale du swollen shoot du cacaoyer sur la production de cacao en milieu paysan à bazré (Côte d'Ivoire). Journal of Applied Biosciences. 2011;**43**: 2947-2957

[5] Kébé BI, Koffie K, N'guessan KF. Le swollen shoot en Côte d'Ivoire: Situation et perspectives. In: Résumés des Actes de la 15ème conférence internationale sur la recherche cacaoyère. (San José, Costa Rica, 9–14 octobre 2006). 2006. p. 66

[6] Kassin KE, Doffangui K, Kouamé B, Yoro G, Assa A. Variabilité pluviométrique et perspectives pour la replantation de la cacaoyère dans le Centre Ouest de la Côte d'Ivoire. Journal of Applied biosciences. 2008;**12**: 633-641

[7] Läderach P. Predicting the Impact of Climate Change on the Cocoa-growing Regions in Ghana and Cote d'ivoire. International Center for Tropical Agriculture (CIAT); 2011. pp. 1-26

[8] Schroth G, Läderach P, Matinez-Valle AI, Bunn C, Jassogne L. Vulnerability to climate change of cocoa in West Africa: patterns, opportunities and limitations to adaptation. Science of the Total Environment. 2016;**556**:231-241. DOI: 10.1016/j.scitotenv.2016.03.024

[9] Adjaloo MK, Oduro W, Banful BK. Floral Phenology of Upper Amazon Cocoa Trees: Implications for Reproduction and Productivity of Cocoa. International Scholarly Research Network; 2012. p. 8

[10] Dje KB. Impacts des phénomènes ENSO sur la pluviométrie et leurs incidences sur la production cacaoyère en Côte d'Ivoire. In: Communication présentée à la Conférence internationale pour la réduction de la vulnérabilité des systèmes naturels économiques et sociaux en Afrique de l'Ouest face aux changements climatiques, Ouagadougou, du 24 au 27 janvier 2007. 2007

[11] Santos E Ad, Almeida AAF d, Branco MCDS, Santos ICD, Ahnert D, Baligar VC. Path analysis of phenotypic traits in young cacao plants under drought conditions. PLoS One. 2018; **13**(2):e0191847. DOI: 10.1371/journal.pone.0191847

[12] Daymond AJ, Hadley P. Differential effects of temperature on fruit development and bean quality of contrasting genotypes of cacao (Theobroma cacao). Annals of Applied Biology. 2008;**153**(2):175-185

[13] Läderach P, Martínez A, Schroth G, Castro N. Predicting the future climatic suitability for cocoa farming of the world's leading producer countries, Ghana and Côte d'Ivoire. Climatic Change. 2013;**2013**(119):841-854. DOI: 10.1007/s10584-013-0774-8

[14] Moser G, Leuschner C, Hertel D, Hölscher D, Köhler M, Leitner D, et al. Response of cocoa trees (*Theobroma cacao*) to a 13-month desiccation period in Sulawesi, Indonesia. Agroforestry Systems. 2010;**79**(2):171-187

[15] CIAT (International Center for Tropical Agriculture). Predicting the Impact of Climate Change on the Cocoa-Growing Regions in Ghana and Cote d'Ivoire. Managua, Nicaragua: CIAT; 2011. Available from: https://www.eenews.net/assets/2011/10/03/document_cw_01.pdf

[16] Braudeau J. In: Larose GPM, editor. Le cacaoyer. Paris; 1969. p. 304

[17] Wessel M, Quist-Wessel PMF. Cocoa production in West Africa, A review and analysis of recent developments. NJAS Wageningen Journal of Life Sciences. 2015;**74–75**:1-7

[18] Parry M, Carter T. Climate Impact and Adaptation Assessment: A Guide to the IPCC Approach. London: Earthscan Publications; 1998. DOI: 10.1002/(SICI)1097-0088(19990330)19:4<459::AID-JOC398>3.0.CO;2-0

[19] Ouranos. Renforcement de la résilience des coopératives des coopératives cacaoyères aux changements climatiques en Côte d'Ivoire – Scénarios et indices climatiques. Rapport présenté par Ouranos et WASCAL-CEA-CCBAD. 2021:53+12

[20] Kalognomou EA, Lennard C, Shongwe ME, Pinto I, Favre A, Kent M, et al. Diagnostic evaluation of precipitation in CORDEX models over Southern Africa. Journal of Climate. 2013;**23**:9477-9506. DOI: 10.1175/JCLI-D-12-00703.1

[21] Maraun D. Bias correcting climate change simulations – A critical review. Current Climate Change Reports. 2016; **2**(4):211-220

[22] Maraun D, Widmann M, Gutiérrez JM, Kotlarski S, Chandler RE, Hertig E, et al. VALUE: A framework to validate downscaling approaches for climate change studies. Earth's Future. 2014;**3**:1-14. DOI: 10.1002/2014EF000259

[23] Kisembe J, Favre A, Dosio A, Lennard C, Sabiiti G, Nimusiima A. Evaluation of rainfall simulations over Uganda in CORDEX regional climate models. 2018. DOI: 10.1007/s00704-018-2643-x

[24] WMO. World Meteorological Organization Guidelines on the Calculation of Climate Normals (WMO-No. 1203). 2017th ed. Geneva; 2017

[25] Daymond A, Tricker PJ, Hadley P. Genotypic variation in photosynthesis in cacao is correlated with stomatal conductance and leaf nitrogen. Biologia Plantarum. 2011;**55**:99-104

[26] Elain SA, Rajesh MK, Balasimha D. Assessment of morphological, physiological and molecular characteristics of cocoa accessions from Central and South America in relation to drought tolerance. Journal of Plantation Crops. 2013;**41**:389-397

[27] Bamba A, Toure NE, Kouadio K, Ahoua SAA, Kouakou DVM, Yoroba F, et al. Contribution of climate scenarios RCP4.5 and RCP8.5 to the study of climate change impacts on cocoa farming

in Côte d'Ivoire; Atmospheric and Climate Sciences. 2023;**13**:84-101. DOI: 10.4236/acs.2023131007

[28] Brunelle R, M'Bo KAA, Okou AK, Tchéma R, Levasseur V. Implementing agroforestry systems in cocoa production as climate change adaptation methods – Case study from Ivory Coast. In: Oral Presentation at the 2022 International Symposium on Cocoa Research (ISCR). Montpellier: France; 2022

Chapter 4

Cocoa Production and Distribution in Bahia (Brazil) after the Witch's Broom

Hélio Rocha Sousa Filho, Marcos de Almeida Bezerra, Raildo Mota de Jesus and Jorge Chiapetti

Abstract

Theobroma cacao production in the state of Bahia (Brazil) suffered crises due to a combination of falling prices, the end of subsidized credit, droughts, international supply and witches' broom disease. The objective was to verify the distribution of the cocoa crop in the state of Bahia and to analyze the indicators of harvested area, production and productivity, starting from the crop crisis that started in the late 1980s. Data were collected from the Brazilian Institute of Geography database. and Statistics, period from 1988 to 2019. Cocoa production is present in 26% of the municipalities, distributed in nine economic regions, especially in the east of the state. Harvested area decreased by 30.7%, production by 65.4% and productivity by 50.1%; numbers that demonstrate the dimension of the problem. In the economic regions, there was a separation of two periods: 1988–1999 and 2000–2019. In the first, the indicators show higher numbers that decrease with the deepening of the crisis. In the second, cultivars resistant to witches' broom and new management and production techniques were implanted, measures related to the behavior of the indicators. Thus, decades after the cocoa farming crisis, increasing production and productivity levels remains a challenge.

Keywords: cocoa cultivation, agricultural production, harvested area, productivity, crisis cocoa

1. Introduction

The cocoa tree (*Theobroma cacao*) is cultivated in tropical areas. Its importance is connected to its seed, the cocoa bean, a commodity traded on international stock exchanges. Driven by the consumption of chocolate-based products, world demand for cocoa beans has been increasing in recent years [1]. In the nineteenth century, most of the cocoa produced in the world came from Latin America, from the tropical zone [2]. In recent years, statistics indicate that Africa has the highest levels of cocoa bean production in the world, followed by Central and South America, Australia, and Asia [3].

In Brazil, cocoa production cannot cover the industry's shortfall in supply, making the country's installed capacity dependent on the importation of cocoa beans [4]. The Brazilian context of insufficient production arose from a combination of factors that contributed to the cocoa production crisis [5]. Witches' broom was one such factor, being a disease of the cocoa tree caused by the *Moniliophthora perniciosa* fungus, one of the most devastating parasites for cocoa production [6]. The fungus infects the plant and leaches nutrients for its growth, in addition to affecting the fruits and causing the loss of beans [7].

After the outbreak of witches' broom on cocoa plantations in South America and the Caribbean, cocoa bean production recorded a fall of 50–90% [8]. In Brazil, witches' broom had previously been recorded in the eighteenth century in the Amazon region, although it was first recorded on cocoa plantations in Northeast Brazil, in the state of Bahia, in 1989 [9].

A crisis began in cocoa production in Bahia at the end of the 1980s, caused by a combination of events such as a drop in prices, the end of subsidized credit, droughts, increased supply from other countries, and fungal diseases. In this scenario, the spread of witches' broom throughout Bahia's cocoa crop aggravated problems in production and productivity, causing unemployment and rural exodus [10]. The fall in production also affected the cities that depended on cocoa cultivation in some way, causing a drop in the circulation of merchandise, decreased municipal tax revenue, and an increase in social problems [5]. One analysis indicated an interdependence between Brazil's cocoa production and that of Bahia, as there was a successive decline in production in Bahia between 1990 and 2004, which impacted national production [11].

Considering the cocoa crisis that began in Bahia in 1989, this study used the harvested area, production, and productivity indicators to investigate the impacts the crisis caused on the cocoa crop. According to the literature, indicators are tools of investigation, measurement, and information on the state of a system, and may be applied to understanding phenomena or processes, managing complex issues, and in decision making [12, 13].

In the context presented above, it is relevant to understand the panorama of cocoa culture in Bahia in the last three decades, after the outbreak of witches' broom, which is one of the factors that most contributed to the decrease in Brazilian cocoa production [10, 11]. On a global level, it is important to gather information on the evolution of cocoa cultivation after an agricultural crisis, especially in Bahia, Brazil, one of the most traditional and most important cocoa-producing regions in the world [14]. The behavior of the indicators may indicate appropriate strategies for crop recovery and the construction of policies that enhance the development of cocoa culture, since positive results in agriculture are drivers of improved quality of rural life [13]. Thus, this study aimed to verify the distribution of cocoa culture in the state of Bahia and analyze the harvested area, production, and productivity indicators, from the crop crisis that began at the end of the 1980s.

2. Material and methods

The cocoa-producing municipalities in the state of Bahia in Northeast Brazil made up the study area. The cultivation of cocoa is of great economic and social importance for Bahia, which is one of the main cocoa producers in Brazil. Economic division of the regions was adopted for the analysis, according to Law no. 6.349 of

17th December 1991, which divided the state into fifteen regions and was adopted by the Superintendência de Estudos Econômicos e Sociais da Bahia (Superintendence of Economic and Social Studies of Bahia) [15]. This division was chosen because it groups the municipalities by economic and commercial characteristics, whereby one or various productive activities identify and determine regional potential [16, 17]. Thus, it may be possible to establish in which regions cocoa culture has the greatest representation.

Information was collected from a database, based on publications by the Instituto Brasileiro de Geografia e Estatística (Brazilian Institute of Geography and Statistics) (IBGE). Information from the Produção Agrícola Municipal (Municipal Agricultural Production) (PAM) survey, which provides data on a municipal, state, and national scale, was also used. The PAM survey is carried out on all production units wholly or partially dedicated to agriculture or livestock. This research is the result of a monthly systematic agricultural survey of annual consolidation [18]. Harvested area, quantity produced (production), and mean yield (productivity) of the permanent cocoa crop in beans, in the municipal and state territories, were selected as indicators. These three indicators were chosen for their availability of access to a chronological series of data. The time frame analyzed was between 1988 and 2019, totaling 32 years. This time frame is from the year before witches' broom was recorded in Bahia up to the last year in which consolidated data are available.

To carry out the analyses, data on harvested area (hectare), production (tons), and productivity (kilogram per hectare) were initially accessed through Table number 1613 of the PAM survey, via the IBGE system of Automatic Recovery. The consolidated values of the variables on state level basis from the database were used in the analysis. Subsequently, the cocoa-producing municipalities in Bahia were selected and classified according to economic region. After this stage, annual values of each of the harvested area, production and productivity indicators of each economic region were calculated, resulting in matrices of 32 × 3, displaying the 32 years observed as rows and the three indicators as columns. The data were auto-scaled and subjected to principal components analysis (PCA). This exploratory method was applied for its capacity to reduce the dimensions of the data and identify inter-relationships between the observations and variables [19]. Execution of the principal components analysis (PCA) will establish patterns of behavior for the indicators, highlighting similarities and/or differences in the evolution of agricultural activity in the regions. Analysis calculations were executed using Past software (Hammer Copyright. 2018, version 3.2, NOR).

3. Results and discussion

The state of Bahia has 417 municipalities, with cocoa production being distributed in 26% of these municipalities (**Table 1**). The data also show growth of around 17% in the quantity of cocoa-producing municipalities when comparing 1988 and 2019. This demonstrates resilience in cocoa cultivation in this period, even with the drop in bean prices, the end of subsidized agricultural credit, and the emergence and spread of witches' broom, which contributed to the third major crisis of the sector [11].

In 1979, the Comissão Executiva do Plano da Lavoura Cacaueira (Executive Commission of the Cocoa Tree Crop Plan) (CEPLAC) carried out a survey on the distribution of the cocoa area in Bahia and its production. In this study, it was found that the cocoa cultivation was present in 83 municipalities, in seven sub-regions, and that some municipalities had the potential to expand their cultivated areas [20].

Region	Municipalities			
	Cocoa producer in 1988	**Cocoa producer in 2019**	**Producer and non-producer in 2019**	**P/T 2019*** **(%)**
Metropolitana de Salvador	1	1	10	10
Litoral Norte	1	1	20	5
Recôncavo Sul	17	20	33	60
Litoral Sul	47	52	53	98
Extremo Sul	13	18	21	85
Sudoeste	14	17	39	43
Nordeste	0	0	47	—
Paraguaçu	2	0	42	—
Baixo Médio São Francisco	0	0	8	—
Piemonte da Diamantina	0	0	24	—
Irecê	0	0	19	—
Chapada Diamantina	0	1	29	3
Serra Geral	0	0	33	—
Médio São Francisco	0	1	16	6
Oeste	0	1	23	4
State total	95	112	417	26

**P/T 2019: percentage of cocoa-producing municipalities in 2019.*

Table 1.
Distribution of the number of cocoa-producing municipalities in the economic regions of Bahia (1988–2019).

According to the most current data from 2019, cocoa production was recorded in nine economic regions in Bahia.

The Litoral Sul and Extremo Sul regions have the highest percentages of cocoa-producing municipalities, with 98% and 85%, respectively. This demonstrates the representativity and importance of cocoa in this part of the state. When examining the history of cocoa in Bahia, it can be noted that a high number of cocoa-producing municipalities are concentrated in these regions. This is not by chance, as cocoa culture began in the south of the state [11]. Another stand-out region is Recôncavo Sul, with cocoa production in more than half of its municipalities, demonstrating that cocoa is also a representative crop of the economy of this region.

The cocoa production record in the municipalities of the Chapada Diamantina, Médio São Francisco, and Oeste regions is relevant data, as they are regions without a tradition of cocoa culture. Trying to make cocoa production viable in these regions may constitute escape zones for cocoa tree diseases, expansion to new production centers, and an opportunity to increase production and generate jobs and income [21]. In addition, it enables crop diversification with the perspective of becoming a new vector of local development.

Analysis of the spatial distribution of the cocoa-producing municipalities in the regions of Bahia, demonstrates that cocoa cultivation is predominant in the east of

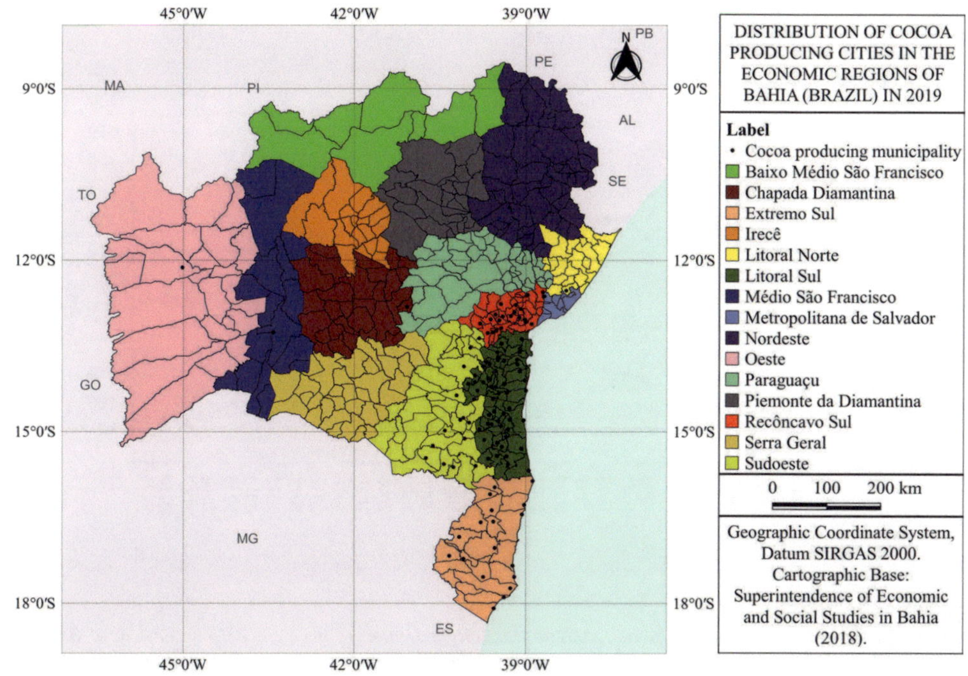

Figure 1.
Location of the 15 economic regions of the state of Bahia and distribution of the cocoa-producing municipalities in 2019.

the state, in the humid zone, in a strip close to the coast (**Figure 1**). This is due to the existence of edaphoclimatic conditions, and processing and exportation infrastructure that favor cocoa production activity [10, 22].

The edaphic and climatic conditions in the south of Bahia are beneficial for cocoa cultivation, as the plant grows in fragments of native forest, in the shade of big trees, with adequate rainfall and deep soils [4, 23]. These conditions contributed to the dissemination of the cocoa tree in the Extremo Sul region, as well as expansion to the municipalities of the humid zone of the Recôncavo Sul and Sudoeste regions.

Starting in 2010, with research and increased technology, experimental areas and cocoa production were created in municipalities in the Chapada Diamantina, Médio São Francisco, and Oeste regions, which do not have naturally favorable edaphoclimatic conditions [24]. Cocoa cultivation was implanted in the humid and sub-humid transition zone and in semi-arid locations. This new expansion is due to irrigated cocoa tree cultivation with sun exposure, the adoption of agronomic management of high technification, and highly productive cloned cultivars resistant to fungal diseases [24].

In recent years, CEPLAC has recommended integrated management in the control of fungal diseases of cocoa trees such as witches' broom, the main pest affecting cocoa plantations in Bahia. The strategies that make up integrated management of witches' broom are culture control, chemical control, biological control, and the insertion of genetically enhanced cultivars [25]. The integrated approach has assisted in production continuity [14]. The large-scale adoption of these strategies may positively interfere in cocoa culture indicators in the producing regions.

Until the mid-1980s, the state of Bahia produced 400 thousand tons of cocoa [14, 26]. Between 1988 and 2019, graphic analysis of the data on the harvested area,

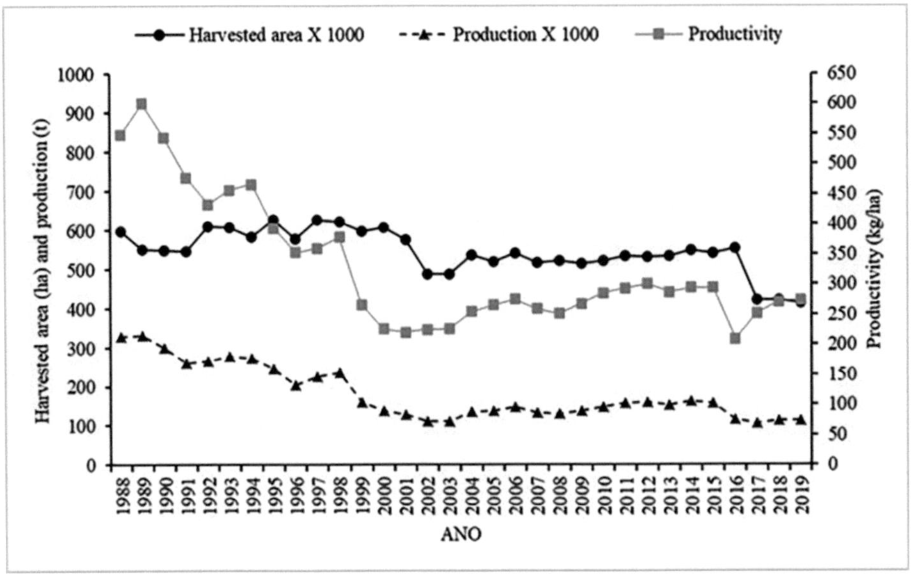

Figure 2.
Evolution of the harvested area, production, and productivity indicators of the cocoa crop in the state of Bahia, between 1988 and 2019. Harvested area (scale of 0–700,000 hectares); production (scale of 0–400,000 tons) and productivity (scale of 0–650 kilograms per hectare).

production, and productivity indicators demonstrates the occurrence of variations and lower final numbers (**Figure 2**). Although witches' broom is indicated as the main cause of these alterations in the 1990s, other factors contributed significantly, such as the fall in the supply of subsidized agricultural credit and the decrease in the price per ton of cocoa beans in the 1980s and 1990s [27].

The harvested area indicator shows a reduction of around 8% in the first four years. From 1992 onwards, there are positive and negative fluctuations, with a general decrease until 2002, when a value of 487 thousand hectares was recorded. From then on, there was an increasing trend until recording 553 thousand hectares in 2016, although this fell value to 413 thousand hectares in 2019. In addition to the fluctuations in harvested area, it decreased by 30.7% between 1988 and 2019.

The production indicator showed a reduction of 66%, between 1988 and 2002. From 2003 onwards, there is a trend of growth until 2015. However, in 2019 production decreased again, recording around 113 thousand tons, which is a fall of 65.4% in relation to 1988. It should be highlighted that despite the reduced production in periods of crisis, cocoa continues to be one of the main agricultural products in the south of Bahia [28].

In the mid-1990s, the decline in cocoa production in Bahia and the need to renew cocoa plantations led the federal government to launch the Cocoa Crop Recovery Program. This program offered credit for producers to invest in recovering plantations and controlling witches' broom. Government technical bodies recommended measures to combat witches' broom disease: phytosanitary pruning, fungicide application, biological control and replanting of cocoa plantations, using clones of resistant cultivars [23].

The productivity indicator showed an increase between 1988 and 1989, from 549 kg/ha to 600 kg/ha. Productivity then decreased to a certain level between 2000 and 2003, recording around 226 kg/ha. Despite oscillations, from 2003 onwards, there was a general trend of increase until 2015. In 2016 production decreased to 209 kg/ha, the lowest recorded yield, this year falling within the period of one of the most severe droughts in Bahia's history, whereby rainfall levels fell below the average, resulting in negative consequences for agriculture and livestock in general [29]. In 2019, productivity reached 274 kg/ha; however, compared with 1988, this is a decrease to the order of 50.1%.

Analysis of the production and productivity indicators shows a positive trend from the beginning of the 2000s, which coincides with the start of the movement substituting traditional cultivars for clones resistant to witches' broom [30]. In the 2000s, CEPLAC started a program of cocoa tree improvement in Bahia that resulted in highly productive cultivars that were resistant to the disease [31].

The cocoa culture problems in Bahia are dealt with on various fronts, with the aim of recovering the losses caused by the crisis. One of the initiatives observed is production for the fine aroma cocoa market, which already has 50 local brands of chocolate in the south of Bahia [10]. This market pays a higher value for cocoa beans in relation to traditional production; however, the production of fine aroma cocoa requires the selection of plant varieties and special care in the ripening, harvest, and post-harvest of the cocoa. This is all done with the aim of achieving distinctive sensory characteristics of aroma and outstanding flavor [32].

The evolution of the cocoa crisis certainly did not affect the harvested area, production, and productivity indicators in the same way or at the same time in all the cocoa-producing regions of Bahia. Thus, through the principal components analysis (PCA), it was possible to observe the interactions between the three indicators as well as the relationships between the observations from 1988 to 2019 (**Figure 3**). Among the ten cocoa-producing regions of Bahia, Chapada Diamantina, Médio São Francisco, and Oeste did not have enough data to be included in the analysis, as production in these regions only began in 2010.

In the Metropolitana de Salvador region, the two PCs (first two components) explain 99.64% of the total variation. PR (production) and PD (productivity) contribute the most to PC1 (first component); HA (harvested area) has the largest contribution to PC2 (second component). The biplot graph for PC1 versus PC2 demonstrates a trend towards separation of 1988 to 1999 from 2000 to 2019. In PC1, the positive vectors of HA, PR, and PD are more related to the period of 1988 to 1999. The results indicate that in most of the period between 1988 and 1999 the indicators were higher (**Figure 3a**).

In the Litoral Norte region, the two PCs explain 99.76% of the total variation (**Figure 3b**). HA and PR contribute the most to PC1; PD has the largest contribution to PC2. The biplot graph for PC1 versus PC2 demonstrates a separation of 1988 to 1999 from 2000 to 2019. In PC1, the positive vectors of the three variables are related to the period of 1988 to 1999, in addition, the lower distance between the HA vector and PR indicates a greater correlation between the two indicators. The data demonstrate that the indicators were higher between 1988 and 1999.

In the Recôncavo Sul region, the two PCs explain 99.67% of the total variation (**Figure 3c**). HA and PR contribute the most to PC1; PD has the largest contribution to PC2. The biplot graph for PC1 versus PC2 demonstrates a separation of 1988 to 1989 from 1990 to 2019. In PC1, the positive vectors PR and PD are more related to 1988 and 1989. The data indicate less variation of the indicators between 1990 and 2019 and higher values.

In the Litoral Sul region, the two PCs explain 99.92% of the total variation (**Figure 3d**). PR and PD contribute the most to PC1; HA has the largest contribution

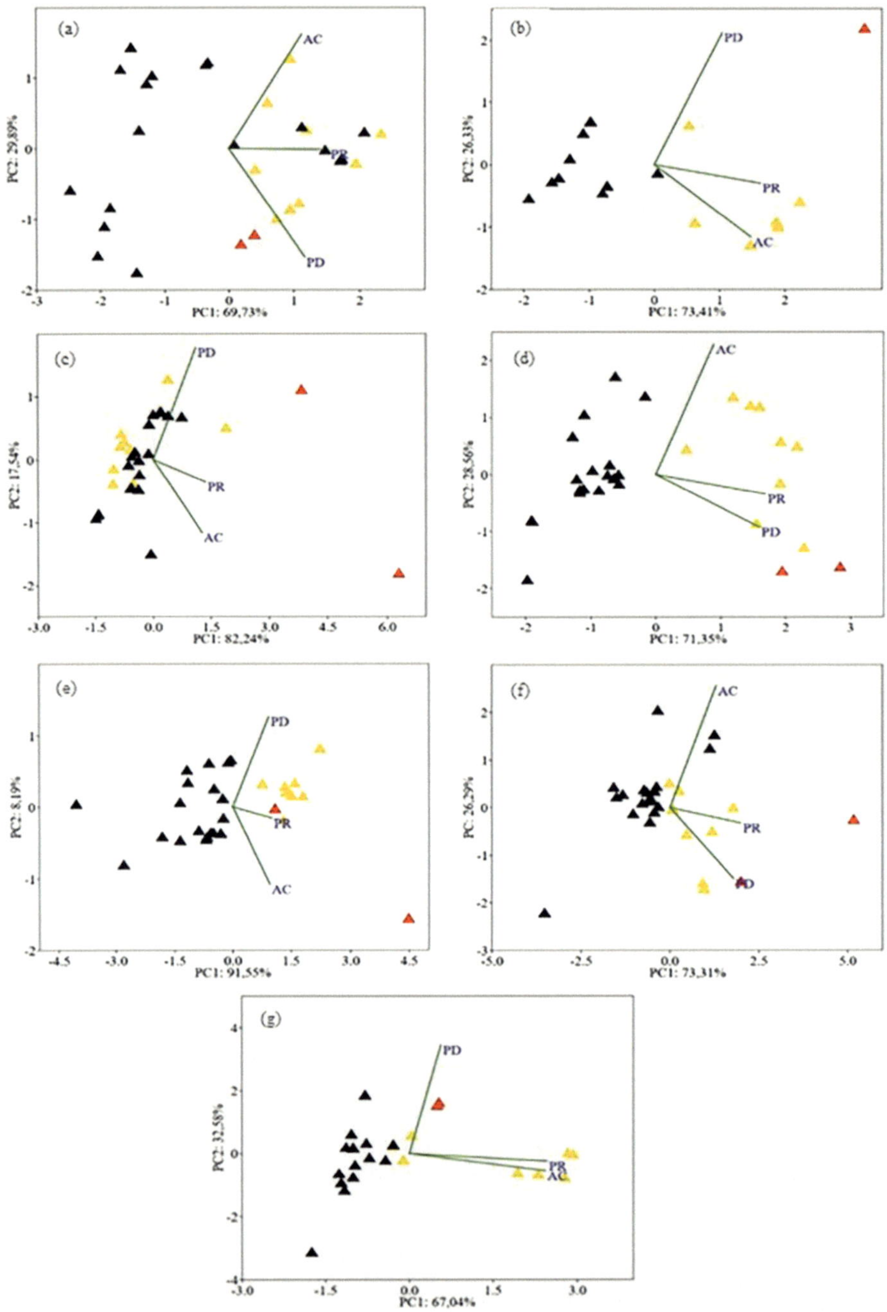

Figure 3.
Principal components analysis carried out on cocoa crop indicators in seven economic regions in the state of Bahia, from 1988 to 2019 (▲1988–1989; ▲1990–1999; and ▲2000–2019). Data from the Produção Agrícola Municipal (Municipal Agricultural Production) survey of the IBGE, using three indicators, HA: harvested area; PR: production; PD: productivity. On the graphs for (a) região Metropolitana de Salvador, (b) Litoral Norte, (c) Recôncavo Sul, (d) Litoral Sul, (e) Extremo Sul, (f) Sudoeste, and (g) Paraguaçu the first two principal components (PC) are demonstrated and explain approximately 95% of the data variability in all the analyses.

to PC2. The biplot graph for PC1 versus PC2 demonstrates a separation of 1988–1999 from 2000 to 2019. In PC1, the positive vectors HA, PR, and PD are related to the period of 1988–1999, in addition, the lower distance between the PR vector and PD indicates a greater correlation between the two indicators. The data demonstrate that the indicators were higher between 1988 and 1999.

In the Extremo Sul region, the two PCs explain 99.71% of the total variation (**Figure 3e**). HA and PR contribute the most to PC1; PD has the largest contribution to PC2. The biplot graph for PC1 versus PC2 demonstrates a separation of 1988–1999 from 2000 to 2019. In PC1, the positive vectors HA, PR, and PD are related to the period of 1988–1999. It can also be seen that from 1990 to 1999 the points are much more clustered, indicating proximity of the values. The data demonstrate that the indicators were higher between 1988 and 1999.

In the Sudoeste region, the two PCs explain 99.64% of the total variation (**Figure 3f**). PR and PD contribute the most to PC1; HA has the largest contribution to PC2. The biplot graph for PC1 versus PC2 demonstrates a trend towards separation of 1988–1999 from 2000 to 2019. In PC1, the positive vectors of PR and PD are more related to the period of 1988–1999, while the HA vector is close to some points from the 2000s. The data demonstrate that in 1988, 1989, and some years from the 1990s, the indicators were higher than the period to 2000–2019.

In the Paraguaçu region the two PCs explain 99.61% of the total variation (**Figure 3g**). HA and PR contribute the most to PC1; PD has the largest contribution to PC2. The biplot graph for PC1 versus PC2 demonstrates a separation of 1988–1999 from 2000 to 2019. In PC1, the positive vectors HA, PR, and PD are more related to the period of 1988–1999, in addition, the lower distance between the HA vector and PR demonstrates a greater correlation between the two indicators. The data demonstrate that the indicators were higher between 1988 and 1999.

In summary, the PCAs carried out on the data on the harvested area, production, and productivity indicators of the cocoa crop, separated or tended to separate the Metropolitana de Salvador, Litoral Norte, Litoral Sul, Extremo Sul, Sudoeste, and Paraguaçu regions into two periods. The first period between 1988 and 1999 is made up of the years following the beginning of the crop crisis, where the values of the indicators are higher and suffer a decrease over the years. In the second period, between 2000 and 2019, the numbers remain lower than in the first period, demonstrating that the productive activity of cocoa did not manage to recover.

In the Recôncavo Sul region, the PCA demonstrates that the data from the 1990s and the 2000s are close (**Figure 3c**), without apparent visual separation. This indicates that in this region, the variation in the harvested area, production, and productivity indicators occurred differently to the other regions, which leads to the inference that the crisis affected the three indicators in the Recôncavo Sul region differently in comparison to the other regions.

Growth rates for the harvested area, production, and productivity indicators were estimated based on the separation of the data into two periods indicated by the principal components analysis. For the harvested area, the growth rate demonstrated that there was a decrease in almost all the regions (**Table 2**). The reduction in the areas with cocoa may be explained by abandonment and substitution of the crop, as with the incidence of witches' broom and all the other previously cited events, maintaining the crop area became a challenge for producers.

The Metropolitana de Salvador, Litoral Sul, and Recôncavo Sul regions were those that had a positive growth rate for the harvested area indicator in at least one period

Region	Harvested area (hectare)					
	Year 1988	Year 1999	GR88/99*	Year 2000	Year 2019	GR00/19* (%)
M. de Salvador	370	684	85%	617	160	–74
L. Norte	250	159	–36%	108	12	–89
R. Sul	56.240	15.597	–72%	15.405	26.266	71
L. Sul	417.052	502.678	21%	509.155	345.131	–32
E. Sul	90.056	53.046	–41%	54.948	27.691	–50
Sudoeste	32.308	25.542	–21%	26.520	13.735	–48
Paraguaçu	113	83	–27%	82	0	–100

GR: growth rate.

Table 2.
Growth rate of the harvested area of the cocoa crop in regions of Bahia (1988–2019).

(**Table 2**). The literature reports that between 1990 and 2004 the advance of witches' broom in 41 municipalities of the Litoral Sul region was not reflected in widespread abandonment or substitution of cocoa for other crops [11]. Regarding the Recôncavo Sul region, it should be pointed out that it is located between humid, sub-humid, and dry climate zones, which may have inhibited the incidence and distribution of witches' broom [23].

The production indicator also had a negative growth rate in most regions. Between 1988 and 1999, the Metropolitana de Salvador region was the only region that presented growth (**Table 3**). In absolute terms, this region has one of the lowest expressions in cocoa production, which may be why this growth has not been considered of great relevance to the general panorama.

Region	Production (tons)					
	Year 1988	Year 1999	GR88/99*	Year 2000	Year 2019	GR00/19*
M. de Salvador	222	386	74%	357	79	–78%
L. Norte	150	55	–63%	33	4	–88%
R. Sul	35.995	6.602	–82%	6.781	9.760	44%
L. Sul	234.035	128.154	–45%	105.326	92.469	–12%
E. Sul	37.115	16.405	–56%	17.124	6.855	–60%
Sudoeste	19.989	7.697	–61%	7.918	3.820	–52%
Paraguaçu	56	29	–48%	29	0	–100%

GR: growth rate.

Table 3.
Growth rate in cocoa tree production in regions of Bahia (1988–2019).

In the Recôncavo Sul region, a positive growth rate is observed between 2000 and 2019 (**Table 3**). A study on cocoa culture in municipalities in this region indicated that the cocoa crisis at the end of the 1980s was not as intense as in the South of Bahia, as, among other reasons, cocoa culture in Recôncavo Sul was intercropped with food crops, differently to the cocoa monoculture in other regions [33]. On this point, it has been suggested that there is a relationship between agricultural crop diversity in a determined region and cushioning of the cocoa crisis.

The Litoral Sul region, the largest cocoa-producing region of Bahia and the most traditional cocoa culture, had a negative growth rate in both periods (**Table 3**). In this region, one of the factors that affected the decrease in production was alteration of the agrarian structure between 1988 and 2017. The cocoa crisis intensified agrarian processes with occupations of farms, the creation of settlements, and some of the old cocoa farms being taken over by squatters [11]. In this context, these small producers were unable, for various reasons, to achieve the level of production of the old farms and maintain the necessary farming practices.

In general, in the 1988/1999 period, all the regions had a negative growth rate for the productivity indicator (**Table 4**). The literature indicates that from the 1990s there was a decrease in productivity of the cocoa crop in Bahia associated with the reduction in agricultural credit, which hampered the use of inputs and farming practices by the farmers [27]. In addition, the witches' broom outbreak at the end of the 1980s also contributed, as it negatively impacts cocoa productivity [23, 34].

In the 2000/2019 period, it was found that only the Litoral Norte and Litoral Sul regions had a positive growth rate for productivity (**Table 4**). Regarding the Litoral Sul region, it should be highlighted that the action of research centers installed in the region may be related to this growth in productivity. CEPLAC, for example, invested in genetic improvement of the cocoa tree in seeking to overcome witches' broom and improve productivity. The renovation of crops with resistant cultivars was implemented and spread among farmers, being indicated as the measure that most created the real possibility of crop recovery [30]. In recent years, superior varieties resistant to witches' broom, if well managed, can reach over 1000 kg/ha/year [35].

Region	Productivity (kg/ha)					
	Year 1988	Year 1999	GR88/99*	Year 2000	Year 2019	GR00/19*
M. de Salvador	600	564	−6%	578	494	−15%
Litoral Norte	600	345	−43%	305	333	9%
Recôncavo Sul	640	423	−34%	440	372	−15%
Litoral Sul	561	254	−55%	206	268	30%
Extremo Sul	412	309	−25%	311	248	−20%
Sudoeste	618	301	−51%	298	278	−7%
Paraguaçu	495	349	−29%	353	0	−100%

**GR: growth rate.*

Table 4.
Productivity growth rate of the cocoa tree crop in regions of Bahia (1988–2019).

4. Conclusions

Up to 2019, cocoa production was present in 26% of the municipalities in the state of Bahia, distributed across 60% of the economic regions. The analysis also demonstrated that between 1988 and 2019, the number of cocoa-producing municipalities increased from 95 to 112, growth of around 17%. These data express the resilience of cocoa culture in Bahia, given the crisis caused by a combination of events at the end of the 1980s.

Examination of the harvested area, production, and productivity indicators indicates that the cocoa culture crisis considerably affected cocoa activity in the state. Comparing 1988–2019, the harvested area decreased 30.7%, production fell 65.4%, and productivity was reduced by 50.1%. These numbers demonstrate the extensive scale of the problem and indicate the need to confront the cocoa crisis on various fronts.

Among the cocoa-producing regions of Bahia, the highest percentages of cocoa-producing municipalities are in Litoral Sul, Extremo Sul, and Recôncavo Sul, with 98%, 85%, and 60%, respectively. This demonstrates the importance of cocoa culture as an agricultural activity that identifies these regions and generates revenue. This merits attention from government bodies connected to agriculture, through decentralized regional development policies and actions. The Chapada Diamantina, Médio São Francisco, and Oeste regions should also be highlighted for beginning cocoa production in areas outside the humid climate zone, which may constitute a promising expansion of cocoa culture with the potential to generate jobs and revenue.

Data analysis on a regional scope also demonstrates that in six economic regions the data separate, or tend to separate, two periods, from 1988 to 1999 and from 2000 to 2019. In the first period, the harvested area, production, and productivity indicators have higher numbers, although they decrease with the deepening of the events that caused the crisis. In the second period, there is the implementation of cultivars resistant to witches' broom and the adoption of new management and production techniques, combat measures that are related to the behavior of the indicators. In the Recôncavo Sul region, the data do not present an apparent separation between the two periods, which indicates that in this region the harvested area, production, and productivity indicators were affected differently by the crisis in comparison to the other regions.

It is concluded that decades after the cocoa crop crisis in Bahia, there remains the challenge of increasing levels of production and productivity, expanding the crop to new regions, and recovering traditional producers.

Acknowledgements

The authors would like to thank the State University of Santa, through the Graduate Program in Development and the Environment, and the State University of Southwest Bahia.

Conflict of interest

The authors declare no conflict of interest.

Author details

Hélio Rocha Sousa Filho[1*], Marcos de Almeida Bezerra[2], Raildo Mota de Jesus[1] and Jorge Chiapetti[1]

1 State University of Santa Cruz, Ilhéus, Brazil

2 State University of Southwest Bahia, Jequié, Brazil

*Address all correspondence to: hrochasousa@gmail.com

References

[1] Beg MS, Ahmad S, Jan K, Bashir K. Status, supply chain and processing of cocoa—A review. Trends in Food Science and Technology. 2017;**66**:108-116. DOI: 10.1016/j.tifs.2017.06.007

[2] Vliet JAV, Giller KE. Mineral nutrition of cocoa: A review. Advances in Agronomy. 2017;**141**:185-270. DOI: 10.1016/bs.agron.2016.10.017

[3] International Cocoa Organization (ICCO). Annual Report 2014/2015. Ivory Coast; 2017. 61 p

[4] Souza CAS, Dias LA d S, Aguilar MAG. Importância Econômica e Social. In: Souza CAS, Dias LA d S, Aguilar MAG, Borém A, editors. Cacau do plantio à colheita. Viçosa: Editora UFV; 2016. pp. 9-40

[5] Chiapetti J. Produção de cacau na Bahia: análise da trajetória política e econômica. In: Souza Júnior JO, editor. Cacau cultivo, pesquisa e inovação. Ilhéus: Editus; 2018. pp. 13-34

[6] Pimenta Neto AA, Laranjeira D, Pires JL, Luz EDMN. Selection of cocoa progenies for resistance to witches' broom. Tropical Plant Pathology. 2018;**43**(5):381-388. DOI: 10.1007/s40858-017-0208-7

[7] Darwiche R, Atab OE, Baroni RM, Teixeira PJPL, Mondego JMC, Pereira GAG, et al. Plant pathogenesis–related proteins of the cacao fungal pathogen *Moniliophthora perniciosa* differ in their lipid-binding specificities. The Journal of Biological Chemistry. 2017;**292**(50):20558-20569. DOI: 10.1074/jbc.m117.811398

[8] Moretti-Almeida G, Thomazella DPT, Pereira GAG, Monteiro G. Heterologous expression of an alternative oxidase from *Moniliophthora perniciosa* in Saccharomyces cerevisiae: Antioxidant function and in vivo platform for the study of new drugs against witches' broom disease. Fungal Genetics and Biology. 2019;**126**:50-55. DOI: 10.1016/j.fgb.2019.02.006

[9] Evans HC, Barreto RW. *Crinipellis perniciosa*: A much investigated but little understood fungus. Mycologist. 1996;**10**:58-61. DOI: 10.1016/S0269-915X(96)80097-8

[10] Dantas PCC, Pires M d M, Uetanabaro APT, Gomes A d S, Pereira AC. O mercado de chocolate no sul da Bahia: estrutura, produção e comercialização. Desenvolvimento Regional em. Debate. 2020;**10**:56-75. DOI: 10.24302/drd.v10i0

[11] Aguiar PCB, Pires M d M. A região cacaueira do sul do estado da Bahia (Brasil): crise e transformação. Cuadernos de geografia de Colômbia. 2019;**28**(1):192-208. DOI: 10.15446/rcdg.v28n1.67437

[12] Olde EM, Moller H, Marchand F, Mcdowell RW, Macleod CJ, Sautier M, et al. When experts disagree: The need to rethink indicator selection for assessing sustainability of agriculture. Environment, Development and Sustainability. 2017;**19**:1327-1342. DOI: 10.1007/s10668-016-9803-x

[13] Cvijanovic D, Stanisic T, Lekovic M, Kostic M. Indicators of agricultural and rural development in the east central and south-east European countries. Agriculture & Forestry. 2020;**66**(2):19-32. DOI: 10.17707/AgricultForest.66.2.02

[14] Evans HC. Withes' broom disease (*Moniliophtora perniciosa*): History

and biology. In: Bailey, Bryan A, Meinhardt LW, editors. Cacao Diseases. Switzerland: Springer; 2016. pp. 137-177

[15] Superintendence of Economic and Social Studies of Bahia (SEI). Economic Regions of the State of Bahia. 2018. Available from: http://www.sei.ba.gov.br/site/geoambientais/mapas/pdf/REGIAO_ECONOMICA_BAHIA_MAPA_2V25M_2018_SEI.pdf

[16] Ortega AC, Cerqueira CA, Silva FPM. As políticas públicas de desenvolvimento no estado da Bahia: evolução e características. In: Ortega AC, Pires MJ d S, editors. As políticas territoriais rurais e a articulação governo federal e estadual: um estudo de caso da Bahia. Brasília: IPEA; 2016. pp. 55-84

[17] Aguiar PCB. Breve panorama econômico e socioterritorial recente da região Cacaueira do Sul do Estado da Bahia, Brasil. Investigaciones Geográficas. 2018;**55**:127-146. DOI: 10.5354/0719-5370.2018.45195

[18] Brasil. Sistema IBGE de Recuperação Automática (SIDRA). By Instituto Brasileiro de Geografia e Estatística (IBGE). 2021. Available from: https://sidra.ibge.gov.br/tabela/1613

[19] Hair JF, Black WC, Badin BJ, Anderson RE, Tatham RL. Análise multivariada de dados. 6nd ed. Porto Alegre: Bookman; 2009. 688 p

[20] Brasil. Aspectos da distribuição da produção e área de cacau do estado da Bahia. By Ministério da Agricultura, Pecuária e Abastecimento (MAPA). 1980. Available from: https://www.gov.br/agricultura/pt-br/assuntos/ceplac/publicacoes/boletins-tecnicos-bahia/bt-075.pdf

[21] Leite JBV, Fonseca EV, Sodré GA, Valle RR, Nascimento MN, Marrocos PCL. Comportamento produtivo de cacau no semiárido do Brasil. Agrotrópica. 2012;**24**(2):85-90

[22] Artuzo FD, Foguesatto CR, Machado JAD, Oliveira L d, Souza ARL d. O potencial produtivo brasileiro: uma análise histórica da produção de milho. Revista em Agronegócio e Meio Ambiente. 2019;**12**(2):515-540. DOI: 10.17765/2176-9168.2019v12n2p515-540

[23] Lima SS, Souza CAS, Patrocínio NGRB, Silva RA d, Santos RSG d, Gramacho KP. Favorabilidade, distribuição e prevalência da vassoura-de-bruxa do cacaueiro no estado do Espírito Santo, Brasil. Agrotrópica. 2018;**30**(1):5-14

[24] Ramos A, Sodré GA, Leite MSB. Irrigação em cacaueiros. In: Souza JJO d, editor. Cacau cultivo, pesquisa e inovação. Ilhéus: Editus; 2018. pp. 437-640

[25] Armengot L, Ferrari L, Milz J, Velásquez F, Hohmann P, Schneider M. Cacao agroforestry systems do not increase pest and disease incidence compared with monocultures under good cultural management practices. Crop Protection. 2020;**130**:105047. DOI: 10.1016/j.cropro.2019.105047

[26] Brasil. Cacau da Bahia: 70% da produção nacional. By Ministério da Agricultura, Pecuária e Abastecimento (MAPA). 2018. Available from: https://www.mda.gov.br/sitemda/noticias/cacau-da-bahia-70-da-produção-nacional.

[27] Chiapetti JA. Crise da atividade cacaueira no contexto do reordenamento da economia mundial. Agrotrópica. 2014;**26**(3):165-174

[28] Chiapetti J, Rocha R B da, Conceição A S da, Baiardi A, Szerman D, Vanwey L. Panorama da Cacauicultura no Território Litoral Sul da Bahia 2015-2019. In: Vanwey, L, Porder S, Chiapetti J, Szerman D, Rocha R B da, Piotto D, editors. Revelando desafios e oportunidades sociais, econômicas, e ecológicas para restauração de florestas em propriedades rurais na Mata Atlântica do Sul da Bahia 2015-2019. Ilhéus: Instituto Floresta Viva: 2020. p. 2-112

[29] Superintendence of Economic and Social Studies of Bahia (SEI). Economic Regions of the State of Bahia. SEI analisa impáctos da seca no estado da Bahia nos anos de 2016 e 2017. 2017. Available from: http://www.sei.ba.gov.br/index.php?option=com_content&view=article&id=2465:sei-analisa-impactos-da-seca-no-estado-da-bahia-nos-anos-de-2016-e-2017

[30] Baiardi A, Ribeiro MCM. Subsídios de política agrícola na cacauicultura dos biomas Mata Atlântica e Amazônia. Revista de POlítica Agrícola. 2018;**27**(2):106-119

[31] Dias LA d S, Neto EF, Souza CAS, Aguilar MAG. Cultivar. In: Souza CAS, Dias LA d S, Aguilar MAG, Borém A, editors. Cacau do plantio à colheita. Viçosas: Editora UFV; 2016. pp. 123-138

[32] Santos AM d, Santos GBM d, Santos PBM d. A estratégia de diferenciação como perspectiva de marcado para o cacau fino. In: Costa FM, Soares NS, editors. Cacau, riqueza dos pobres. Ilhéus: Editus; 2016. pp. 219-237

[33] Almeida LS d. O Vale do Jiquiriçá no contexto do circuito espacial produtivo do cacao (thesis). Salvador: Universidade Federal da Bahia; 2008

[34] Moda LR, Boteon M, Ribeiro RG. Cenário econômico do mercado de cacau e chocolate: oportunidades para cacauicultura brasileira. Brazilian Journal of Development. 2019;**5**(10):21203-21225. DOI: 10.34117/bjdv5n10-281

[35] Ahnert D, Melo HL, Santos FFJ, Lima LR, Baligar VC. Melhoramento genético e produtividade do cacaueiro no Brasil. In: Souza Júnior JO, editor. Cacau cultivo, pesquisa e inovação. Ilhéus: Editus; 2016. pp. 151-182

Chapter 5

Challenges and Opportunities for Indonesian Cocoa Development in the Era of Climate Change

Kurnia Dewi Sasmita, Edi Wardiana, Saefudin, Dibyo Pranowo, Asif Aunillah, Nur Kholilatul Izzah, Maman Herman, Nur Kholis Firdaus, Iing Sobari, Sakiroh and Dewi Listyati

Abstract

In recent years, the area of cocoa plantations in Indonesia has tended to decline, one of which is attributed to climate change that threatens the sustainability of production; even though cocoa production and consumption have become popular globally, the consumer demand for cocoa products has also increased. Climate change causes increased air temperature, erratic rainfall patterns, increased sea level and surface temperature, and extreme weather. Cocoa requires an ideal rainfall of 1500–2500 mm/year and dry months (rainfall <60 mm) for about 1–3 month a year. Climate change can be a challenge for Indonesian cocoa development. Several efforts should be made to turn existing challenges into opportunities through appropriate technological inputs, such as the use of improved cocoa genetic resources (recommended clones) as well as improving nursery and field management practices, including shading and watering the seedlings, modification of growing media, mycorrhizal application, rainwater harvesting, and managing shade plants and intercropping.

Keywords: cacao clone, growing media, harvesting rainwater, intercropping, mycorrhiza, shading, watering

1. Introduction

Indonesia is one of the world's largest cocoa producers. Most of Indonesia's cocoa (about 80%) is cultivated on smallholder plantations (SP), while a small portion is owned by the state (government plantations (GP)) and private plantations (PP). Over the past 5 years, Indonesia's cocoa bean production has decreased from 270,000 tons in 2017 to 170,000 tons in 2021. This has resulted in Indonesia's ranking as the world's fourth largest cocoa producer dropping to seventh after Côte d'Ivoire, Ghana, Ecuador, Cameroon, Nigeria, and Brazil. One of the main causes is the global climate change [1].

 IntechOpen

Based on data [2], in the last 10 years (2013–2022), Indonesian cocoa has faced the problem of decreasing planting area. The average rate of decline in the area of Indonesian cocoa plantations has reached 1.80%, and based on the plantation conditions, immature young plants have decreased drastically, around 6.62% per year. Even though the reduction in the area does not have much impact on national cocoa production, this trend will pose a significant threat to the development of Indonesian cocoa in the future. However, during 2013–2022 there was a slight increase in production (0.96% per year) due to an increase in harvested area, although the productivity remained below 1 ton/hectare/year (**Figure 1**).

Many factors affect the decline in the area and productivity of Indonesian cocoa, one of which is due to the impact of climate change. The effect of climate change on cocoa growth and production is very significant because cocoa cultivation is highly dependent on climatic conditions. This impact affects not only Indonesian cocoa but almost all cocoa plantations in the world. Previous studies in Ghana and Côte d'Ivoire showed that due to the impact of climate change, there would be a significant reduction in the extent or size of suitable lands for cocoa for estimates until 2050 [4]. The impact of the 2015–2016 El Niño on cocoa in Brazil caused high plant mortality and a severe decline in production [5]. Therefore, the impact of climate change on cocoa has become a global concern; on the one hand, world cocoa production tends to decrease, while on the other hand, consumer demand continues to increase.

Climate change causes an increase in air temperature, erratic rainfall patterns, an increase in sea surface temperature and sea levels, and extreme weather. These changes affect not only environmental factors related to plant physiological aspects [6] but also the aspects of the diversity and spread of pests and diseases. Climate change is an important factor affecting the quality of cocoa beans. A long dry season leads to a smaller size of cocoa beans, and the flowers easily wilt or fall off [7], and an increase in temperature in the cocoa planting area accelerates the fruit ripening process, resulting in imperfect flavors [8]. In addition, climate change affects pests' life

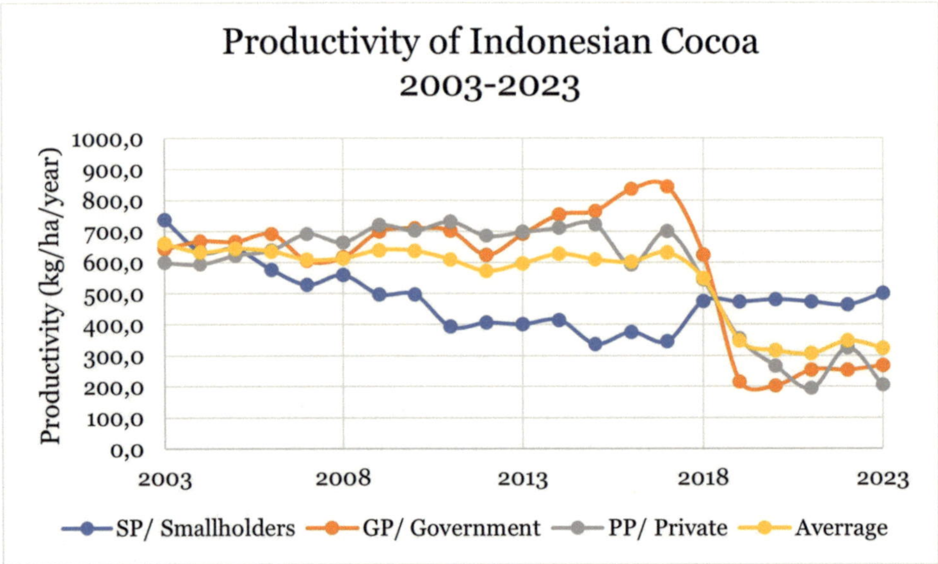

Figure 1.
Productivity of Indonesian cocoa, 2003–2023. Source: [3].

cycles, which indirectly affect the growth and production of cocoa. The intensity of cocoa pod disease attacks increases due to climate change [9]. In Brazil, the spread of the fungus *Phytophthora palmivora*, which causes black pod disease in cocoa, increases during the more intense rainy season and higher temperatures [10]. Moreover, the biological agents controlling cocoa pests are also affected, so the pests become resistant and attack cocoa plants easily [11]. High rainfall will also influence production due to low solar radiation, thereby reducing the number of flowers and delaying fruit ripening [12]. Therefore, in addition to the need for adaptation through good plant management practices, breeding programs for resistance to biotic and abiotic factors also play an important role [6, 11].

In Indonesian cocoa plantations, the adaptation process to climate changes can be carried out in various ways, including the extensification and intensification programs. However, the options for area extensification programs are rather limited, considering the increasing competition for land use with other sectors. The experience in Ghana and Côte d'Ivoire showed that the optimal altitude for cocoa cultivation is estimated to increase toward the hilly areas. However, this area is included in the protected forest zone. Thus, it will be an obstacle if used for cocoa expansion [13].

This chapter identified the strategy for adaptation of cocoa plantations to climate change and emphasized intensification by improving plant genetic resources and managing plants from the nursery to the field level. In addition, fixing the limitations and deficiencies of land suitability needs to be carried out through appropriate technological inputs.

2. Challenges facing Indonesian cocoa in the era of climate change

2.1 Impact of climate change on the ecosystems of cocoa plantation

Climate change has brought about changes in several climatic elements, including increasing air temperature, changing rainfall patterns, increasing sea surface temperature and sea levels, and extreme weather. Climate changes affect the agricultural ecosystem, including the cocoa ecosystem. Climate changes also impact the degradation of soil quality (physical, chemical, and biological), so it could disturb plant growth, and even many cocoa plants have died. Such conditions contribute to a decline in the number (population) of cocoa plants in Indonesia. Hence, efforts are needed to reduce the negative impacts of climate change through appropriate technological innovations.

2.1.1 Increasing air temperature

Based on the studies, the temperature of the earth's surface was predicted to increase by 2.1–3.9°C due to global climate change [14]. The measurements over 95 years (1901–1995), in the low latitudes of Asia, Africa, and Central America, air temperature increased with variations of less than 1.5°C, while at the higher latitude, their increase was relatively greater [15]. In various regions of Indonesia, there is also an increase in air temperature, which varies from one region to another. Over the past 30 years (1991–2020), temperatures in the Indonesian regions have increased by an average of around 0.9°C [16].

The increase in air temperature is related to the changes in relative humidity and the evapotranspiration rate in agricultural areas, including cocoa plantations. If the air temperature increases, evapotranspiration increases while the relative humidity decreases. In the dry season, the changes are more significant because an increase in solar radiation triggers them. On the other hand, there is a substantial difference between the temperature of the day and the night. Previous research on cocoa plantations in East Java showed that the increase in temperature within 2–4 months before harvest time had a negative impact on production [17]. Similarly, in Gunungkidul, Yogyakarta, the changes in air temperature are one of the factors affecting fluctuations in cocoa yields [18].

2.1.2 Extreme weather events: Changes in rainfall, increasing temperatures, and rise in sea levels

One of the effects of climate change is that rainfall patterns become irregular, making the wet and dry seasons more erratic. At the same time, one area may experience drought, while another area will experience flooding and landslides. Prolonged drought will cause forest fires and trigger soil degradation, reducing soil fertility and killing various soil microorganisms. The final impact of these problems will result in reduced agricultural land that is considered suitable for certain commodities, including areas for cocoa cultivation.

Cocoa plants only require dry months (rainfall <600 mm/month) for 1–3 months. Therefore, the occurrence of prolonged dry months due to the impact of climate change will reduce cocoa production and quality. Dry months for 1–6 months before harvest reduce the cocoa pod quality in Sukabumi, West Java [19]. Meanwhile, decreased rainfall 2–4 months before harvest has negatively impacted cocoa production in East Java [17]. In 2019, a prolonged drought in Lampung reduced the number of cocoa pods, even causing crop failure [20]. Drought stress reduces vegetative growth, plant biomass, relative water content, leaf chlorophyll content, and nitrate reductase activity, while leaf phenol and proline content increased [21]. Drought stress also reduces mesophyll thickness, leaves, and abaxial epidermis, affecting the stomata's density and closure [22]. Other studies have shown that the number of dry months was correlated positively with the severity of vascular streak dieback (VSD) disease [23].

On the other hand, excessive rainfall causes soil erosion, flooding, and landslides. Soil erosion causes a decrease in soil fertility, which reduces the productivity of cocoa plants. Floods will damage cocoa plantations at low and medium altitudes. In addition, the landslide risk can occur because cocoa is generally intensively cultivated in low to medium altitudes (0–700 m above sea level (asl)). In addition to heavy rainfall, flooding will damage cocoa plantations in the coastal areas due to rising sea levels caused by climate change. During excessive rainfall, the spread of cocoa disease caused by *Phytophthora palmivora* and *Ceratobasidium theobromae* is relatively faster. The rainfall factor causes fluctuations in cocoa yields in Gunungkidul, Yogyakarta [18].

Another impact of climate change is extreme weather, which causes heavy rains, strong winds, and high waves in water areas. The indirect effect of extreme weather is reduced biodiversity, such as pollinators, birds as seed dispersers, and populations of natural enemies for pests and diseases [23]. The effect of La Niña in 2016 reduced the number of cocoa pods in Bali, and the vegetative growth of plants appears to be more dominant than the generative growth [24].

2.2 Development of Indonesia's cocoa area and production, 2013-2022

The total area of Indonesian cocoa during the last 10 years (2013–2022) has decreased by 1.80% per year. The sharp decline occurred in government plantations

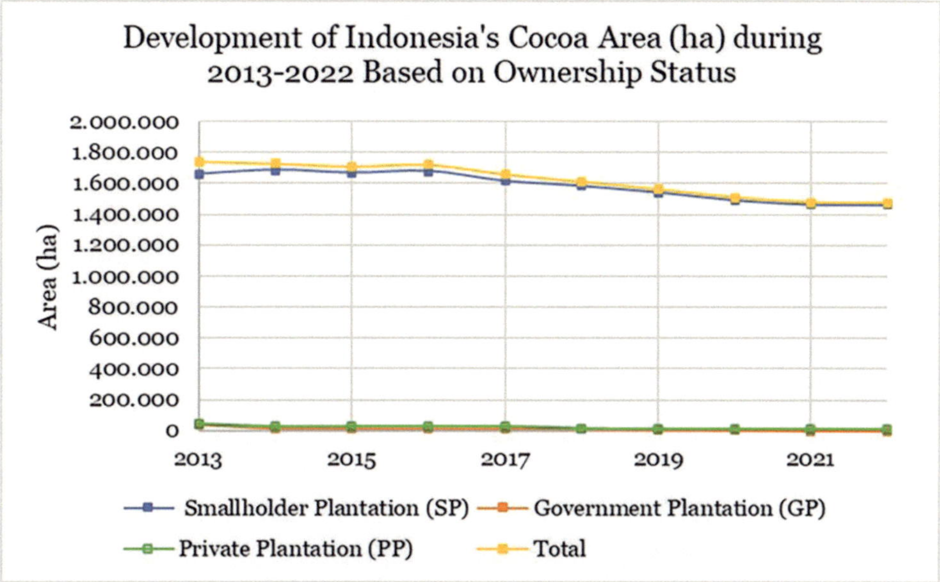

Figure 2.
Development of Indonesia's cocoa area (ha) during 2013–2022 based on ownership status. Source: [2].

(GP) and private plantation (PP) by 28.33 and 11.33% per year, respectively. Smallholder plantation (SP), which dominates Indonesia's cocoa population, has experienced a decrease in land area but is relatively smaller than those of GP and PP, which is around 1.37% per year (**Figure 2**).

Figure 3 shows that the area of mature plants (MP) increased by 1.99% per year. In line with the increasing plant age, many plants changed their status. Initially they

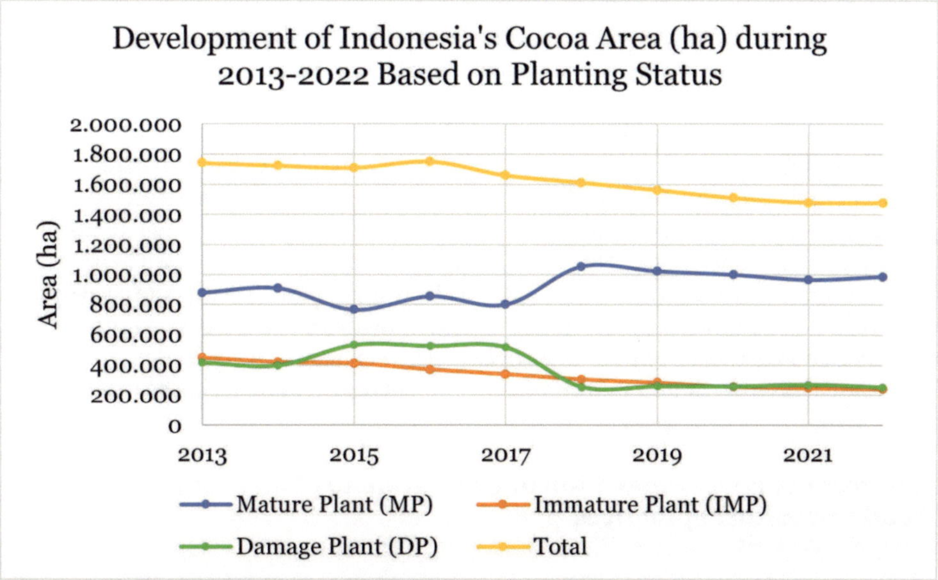

Figure 3.
Development of Indonesia's cocoa area (ha) during 2013–2022 based on planting status. Source: [2].

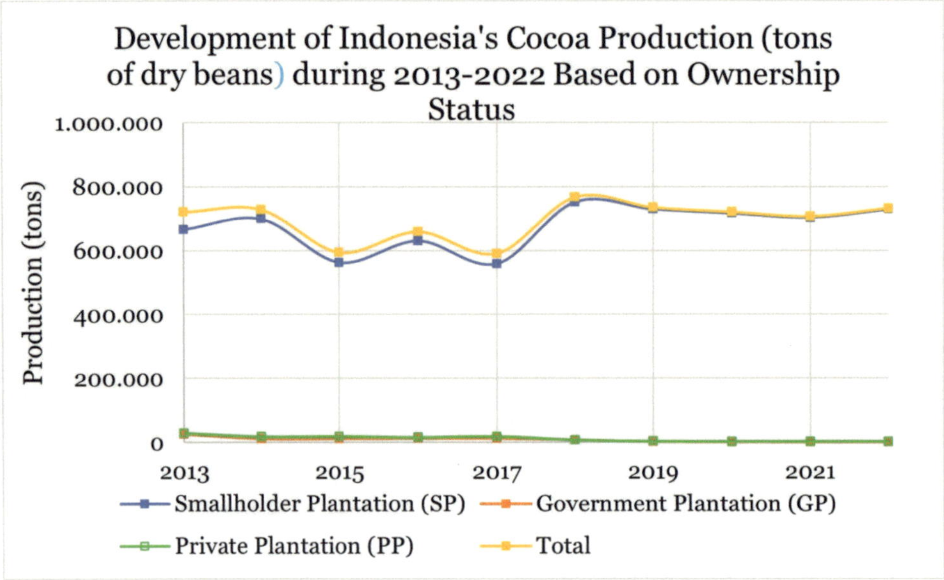

Figure 4.
Development of Indonesia's cocoa production (tons of dry beans)) during 2013–2022 based on ownership status. Source: [2].

were categorized as immature plantations (IMP), then changed to MP. However, the increase in MP number was not proportional to the decrease in IMP (about 6.62% per year), even though the damaged plant (DP) decreased by 2.71% per year. It indicates that during the last 10 years (2013–2022), many young cocoa plants have died before entering their production period. The declined area of DP is caused by the death of many plants, causing a decrease in the total area. This phenomenon will threaten national cocoa production in the next few years.

Many factors contribute to the high rate of decline in IMP, such as stress on the biophysical environment due to changes in climatic conditions. Cocoa plants that are stressed by the biophysical environment will be very vulnerable to pests and disease attacks, even though control has been carried out. Therefore, it is necessary to take action to suppress and reduce the impact of climate change by manipulating growing environmental conditions and implementing good agricultural practices.

Indonesia's total cocoa production during 2013–2022 increased, although relatively small, by about 0.96% per year (**Figure 4**). The main driver of this increase in production was the increase of MP area, as shown in Table 2. Over the past 10 years, the cocoa production of SP has increased by 1.98% per year. Meanwhile, GP and PP cocoa production experienced a sharp decline of 30.49 and 17.6% per year, respectively. The extent of the SP planting area has contributed significantly to the increase in national cocoa production.

3. The technological innovation of adaptation to climate change through good agricultural practices

Cocoa crop management in the era of climate change requires several adaptation strategies, including the use of drought-tolerant clones, followed by improvements in cultivation management from the nursery to the field level.

3.1 Drought-tolerant clones

The use of drought-tolerant clones is the leading strategy in dealing with the phenomenon of climate change. Cocoa clones known to be drought-resistant are Sca 6, Amelonado, TSH 919 [25], KW 163, KW 165, KW 215 [19], KW 641 [23, 26], KW 514, KW 535, KW 619, KW 516 [26], and KW 562 [27].

3.2 Management at nursery level

3.2.1 Shading and watering of cocoa seedlings

Several methods can be used in water-saving irrigation systems in nurseries, such as sprinkler irrigation, drip irrigation, and capillary wick system. Shading of cocoa seedlings using "paranet" made from nylon materials and watering using sprinkler irrigation method at the Pakuwon Agroscience Park, Indonesian Agricultural Research and Development Agency, has been proven to improve seedling growth and save water use (**Figure 5**). Dealing with water shortages for nurseries, especially during the dry season, can be assisted by micro and semipermanent rainwater harvesting ponds (**Figure 6**).

3.2.2 Modification of growing media

Modifying the growing media has been shown to increase the resistance of cocoa seedlings under water-stress conditions. Several ameliorants can be used as growing media for cocoa seedlings, including manure, compost, biochar, sawdust, and hydrogel [28–30]. Using up to 9 g of cocoa pod husk (CPH) biochar per kg of soil and 12 g

Figure 5.
Shading and sprinkler irrigation for cocoa and coffee seedling at Pakuwon Agroscience Park, in Sukabumi, West Java.

Figure 6.
Micro and semipermanent rainwater harvesting ponds at Pakuwon Agroscience Park, in Sukabumi, West Java.

of CPH biochar per kg of soil with every 6 days of water frequency increases water use efficiency by 208.8 and 262.22%, respectively, compared to control (without biochar) [28]. In addition, adding compost to growing media increases the height, stem diameter, number of leaves, and root weight of cocoa seedlings [31]. Growing media mixed with sawdust, rice husk biochar, and compost in ratios of 60:20:20 and 60:10:30 recovered cocoa seedlings after water stress [30].

3.2.3 Mycorrhizal applications

Mycorrhizal inoculation in nurseries is crucial because it overcomes the problems of limited nutrition, various abiotic stresses (salinity, drought, extreme temperatures, and acidity), and biotic stresses. Therefore, it is necessary for seedlings to be resistant to environmental stress and soil-borne pathogens [32]. Application of mycorrhiza increases water use efficiency [33, 34], nutrient uptakes [34], and seedling growth of cocoa [34–39]. Mycorrhizal inoculation at the nursery stage is expected to have a positive impact on the growth of cocoa plants in the field.

Research progress (1):

An experiment entitled "The Application of Mycorrhiza and Ameliorants to Improve the Growth of Cocoa Plants in Acid Soil" was conducted in a greenhouse at the Pakuwon Agroscience Park, Sukabumi, West Java, starting in 2022/2023. Cocoa seedlings were planted in the polybags on acidic soil media. A completely randomized factorial design consisting of two factors with three replications was used in this study. The first factor consisted of mycorrhizal and without mycorrhizal applications, while the second factor was the application of ameliorant (dolomite and cow manure) and control. Preliminary results showed that mycorrhizal application in acid soil increases cocoa seedlings' root volume and fresh weight. In addition, cow manure had a better effect than dolomite and control (with or without mycorrhizal application) (**Figures** 7 and 8).

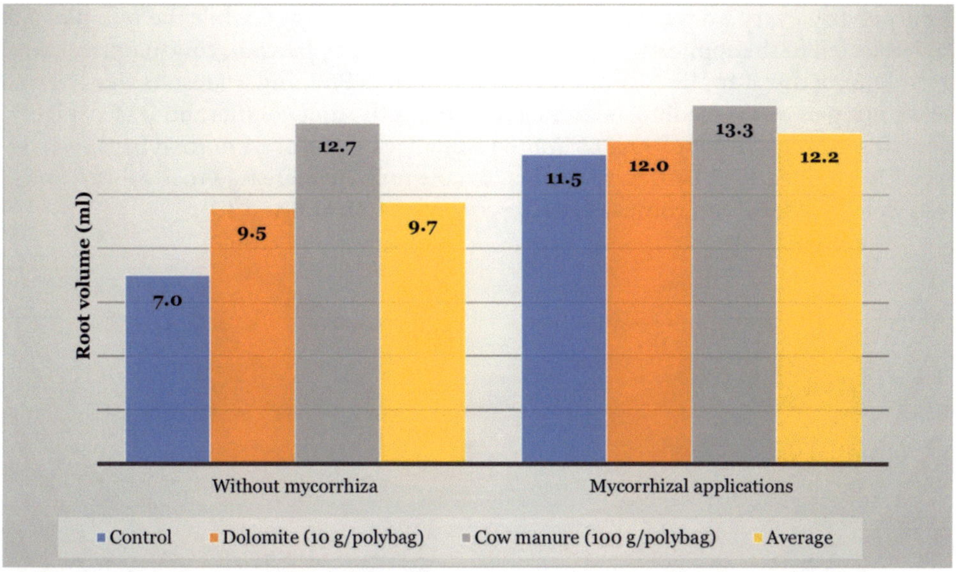

Figure 7.
The effect of mycorrhizal, dolomite, and cow manure on the root volume of cocoa seedlings at 3 months after application in acid soil.

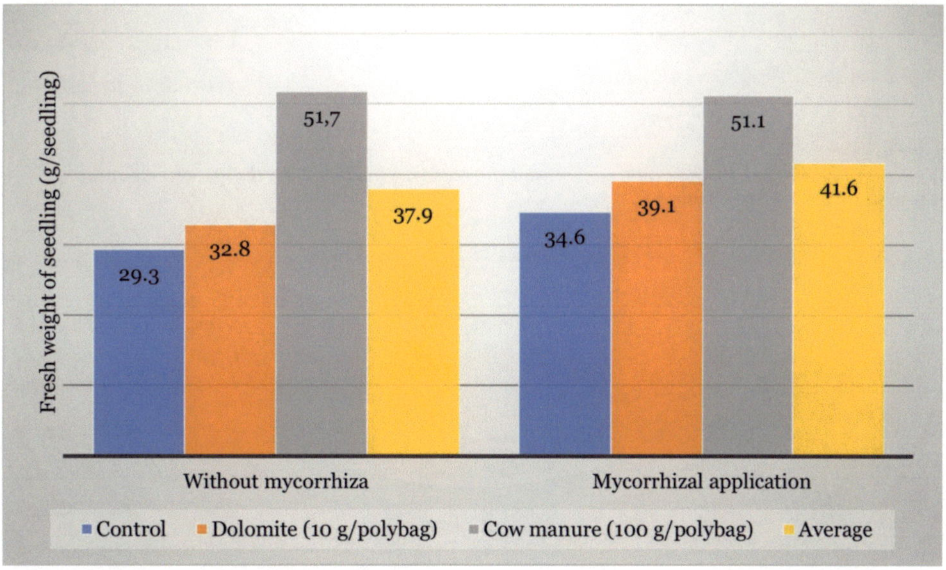

Figure 8.
The effect of mycorrhizal, dolomite, and cow manure on the fresh weight of cocoa seedlings at 3 months after application in acid soil.

Research progress (2):

The experiment entitled "The Application of Mycorrhiza and Ameliorants to Improve the Growth of Cocoa Plants in Drought Conditions" was conducted in a greenhouse at the Pakuwon Agroscience Park, Sukabumi, West Java, starting in 2022/2023. Cocoa seedlings were planted in the polybags. A split plot design with four replications was used in this study. The main plot consisted of two watering treatments, 123 and

61 ml/polybag every 3 days (applied 2.5 months after planting), while the split plot was subjected to the application of mycorrhizal, hydrogel, biochar, cow manure, and control mycorrhizal applied 1,5 months after planting. Preliminary results showed that the watering of cocoa seedlings significantly affects the root volume and fresh weight of cocoa seedlings. In conditions of limited water (watering of 61 ml/polybag), using mycorrhiza added with hydrogel, biochar, or cow manure increased root volume and fresh weight of seedlings compared to the control (**Figures 9** and **10**).

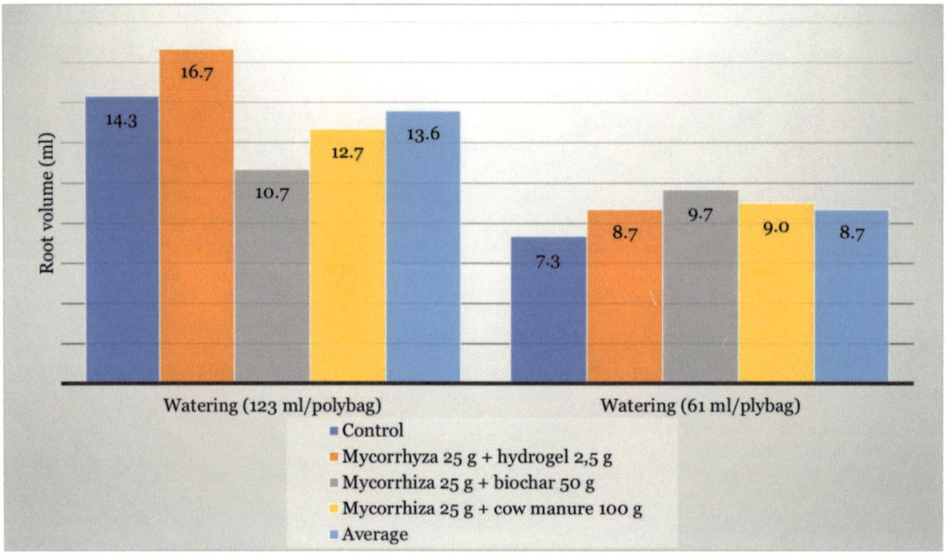

Figure 9.
The effect of mycorrhizal, biochar, cow manure, and two watering treatments on the root volume of cocoa seedlings at 5 months after planting.

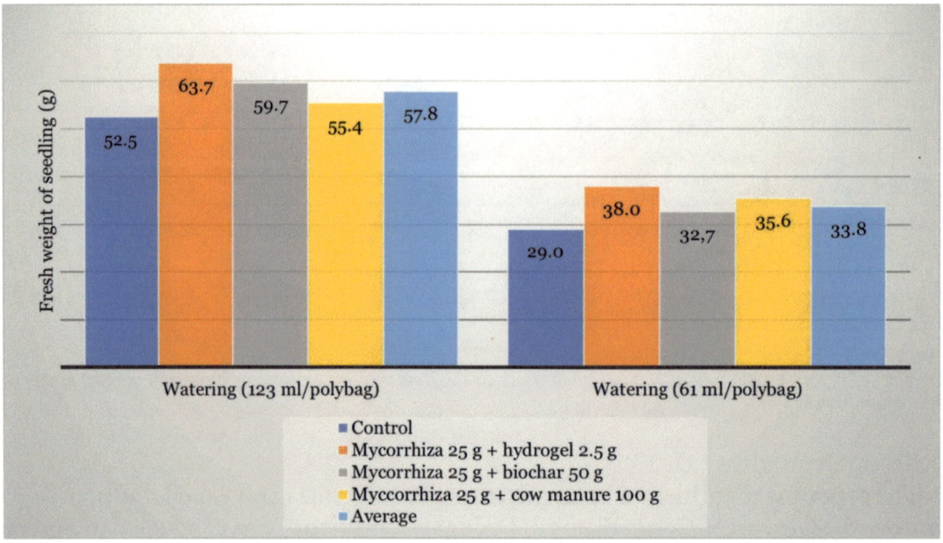

Figure 10.
The effect of mycorrhizal, biochar, cow manure, and two watering treatments on the fresh weight of cocoa seedlings at 5 months after planting.

3.3 Crop management at the field level

Crop management at the field level is focused on three main activities, such as: (i) harvesting rainwater by constructing water storage ponds, (ii) shading cocoa to reduce the effect of solar radiation and high temperatures, and (iii) reducing the evapotranspiration rates by planting intercrops or cover crops.

3.3.1 Harvesting rainwater

The technology for rainwater harvesting is through making ponds, channel reservoirs, and "rorak" systems (sediment pits). Making ponds for cocoa planting is more directed to meet water needs at the start of planting when cocoa's sensitivity to water is limited. The ponds are simple in their construction and size, so it applies to the farmer's level (**Figure 11**). In large cocoa planting areas, permanent ponds of a larger size are needed so that they can serve the existing cocoa population. In this case, the role of the government is very much needed considering the cost of building such a reservoir is very high. Meanwhile, during the dry season when the water starts to decrease, the rorak can function to accommodate organic matter sourced from the pruning of cocoa plants, shading plants, or cover crops. Mowidu and Endang Sri Dewi [40] stated that using rorak in cocoa cultivation increases the total nitrogen (N) and phosphorus (P) available in the soil.

3.3.2 Shading cocoa

Cocoa is a shade-demanding crop that requires shade crops, with only about 60–80% of the solar radiation necessary for growth. Without shade crops, cocoa yields can be increased in the short term, but in the long term, the sustainability of

Figure 11.
The simple ponds for harvesting rainwater at smallholder cocoa plantation, Soppeng, South Sulawesi.

production is threatened. Physiologically, the function of shade on cocoa plants is related to the improvement of microclimate and site conditions (reduction of air and soil temperature extremes, reduction of wind speed, maintenance of soil moisture and water availability, improvement and maintenance of soil fertility including reduction of erosion) and the reduction of solar radiation penetration in both quantity and quality to avoid flushing in cocoa [41]. Therefore, in the current era of climate change, the discussion on the function and role of shade trees for cocoa crops has become a matter of great urgency. Improper management of shade trees will not provide optimal benefits. In fact, it will have a negative impact on cocoa plants. Proper management of shade trees can support growth and increase yields. However, too much shade creates a microclimate that favors the incidence of diseases. Therefore, the selected shade trees should have an architecture capable of transmitting optimal solar radiation for cocoa, their roots should not compete with cocoa for water and nutrients, they should not become host for cocoa pests and diseases, they should be relatively easy to plant and maintain, and they should have a high economic value.

The critical period for cocoa is the young plant stage or initial planting period. Therefore, shade plants should be planted before cocoa, so they can provide shade when cocoa begins to enter production period. In the early stages of cocoa growth, temporary shade plants can use annual or biennial crops such as bananas (**Figure 12**). Bananas have wide canopies that could shade young cocoa plants and maintain adequate soil moisture levels. And for permanent shade perennial crops, such as coconut, areca nut, rubber, *Gliricidia, Albizia*, *Leucaena, Cassia,* or *Erythrina*, can be used. The

Figure 12.
Bananas as temporary shade plants for young cacao at smallholder cocoa plantation, Soppeng, South Sulawesi.

Figure 13.
Mulch from cocoa pruning at cocoa plantations of Pakuwon Agroscience Park, in Sukabumi, West Java.

research results at cocoa plantations in Sulawesi with *Gliricidia* shade plants showed their tolerance to drought [42, 43].

The increase in air temperature due to climate change requires shade trees to be planted at closer distance. Pruning of shade trees is carried out according to seasonal conditions, where pruning should not be carried out in the dry season to allow the tree to withstand too much sunlight. Pruning of shade trees should be done before the rainy season. The leaves from cocoa pruning can be utilized as mulch to minimize evapotranspiration and as a source of organic matter (**Figure 13**).

3.3.3 Intercropping

Cocoa planted under shade plants is an intercropping pattern. Cocoa is planted between two rows of shade crops, and the distance between the cocoa and shade crops is quite large. This cropping pattern still leaves space between the two crops, so there is still an opportunity to plant other crops as intercrops. Intercropping between cocoa and shade trees suppresses weed growth, reduces evapotranspiration, and can withstand runoff, erosion, and landslides on sloping topography. In addition, planting intercrops will produce various agricultural products, so if properly and correctly managed, it will sustainably increase farmers' income. By planting intercrops on an annual or a biennial basis, results will be obtained quickly, the frequency of tillage will be more intensive, so that the soil's physical, chemical, and biological conditions will always be maintained, and as a source of easy and cheap organic material.

The integration of cacao, shade plants, and intercrops is similar to the cacao agroforestry model that has been widely discussed in the era of climate change. A dynamic cocoa agroforestry system can develop various crops with different life cycles. It resembles multistoreyed cropping systems [44] and high-density cropping systems in coconut [45].

The selection of intercrops to be planted between cacao is based on the microclimatic conditions between the canopies and market opportunities. Some intercrops identified as tolerant and able to adapt well to highly shaded conditions are colocasia, amorphophallus, banana, ginger, turmeric, arrowroot, and kacholam [46]. In addition, the selection of intercrops should also consider the habitus of the trees lower than cacao, the rooting structure of the intercrops should be different from that of cacao, and intercrops should not be hosts to cacao pests and diseases. On the other hand, legume cover crops as intercrops may not provide direct economic returns, but these crops can help fertilize the soil, suppress weed growth, and retain runoff to prevent sloping topography from erosion and landslides. In livestock areas, planting fodder crops as intercrops can easily and sustainably provide fodder, allowing for mixed farming. Research shows that sequential planting of rice and soybean as intercrops under cocoa could improve productivity in the long term [47], as can intercropping with patchouli; even the nutrients given to patchouli plants can be absorbed by cocoa plants, allowing them to grow better than cocoa monoculture [48]. In addition to choosing the right type of intercrops and spacing them appropriately, providing adequate intakes for each combined crop (cocoa, shade plants, and intercrops) will avoid the effects of water and nutrient competition that may occur.

4. Conclusion

The impact of climate change on cocoa production in Indonesia has been highlighted. The rate of decline in Indonesia's cocoa area during 2003–2022 is 1.80% per year, with young cocoa plants declining the most at 6.62% per year. During the same period, the slight increase in production (0.96% per year) is not due to increased productivity but rather to an increase in harvested area. As a result, Indonesia has fallen from fourth to seventh place as the world's largest cocoa producer. To minimize further impacts of climate change on Indonesian cocoa, adaptation strategies through improved crop management are needed. Strategies through extensification programs are still open to opportunities, but these strategies face obstacles because they will compete with other sector development. The intensification strategy that needs to be implemented starts from selecting cocoa clones that are considered drought tolerant, followed by good nursery management through watering and shading, modification of growth media, and application of mycorrhiza to produce planting material with good vigor. Furthermore, at the field level, practices, such as rainwater harvesting technology, optimal management of shade crops, and planting appropriate intercrops, should be promoted for adoption by farmers for sustainable cocoa production. The advantage of yields of intercrops can meet farmers' short-term food and income needs.

Author details

Kurnia Dewi Sasmita[1*], Edi Wardiana[1], Saefudin[1], Dibyo Pranowo[1], Asif Aunillah[2], Nur Kholilatul Izzah[1], Maman Herman[1], Nur Kholis Firdaus[1], Iing Sobari[1], Sakiroh[1] and Dewi Listyati[3]

1 Research Center for Horticultural and Estate Crops, National Research and Innovation Agency, Cibinong, Bogor, Indonesia

2 Research Center for Agroindustry, National Research and Innovation Agency, Bogor, Indonesia

3 Research Center for Behavioral and Circular Economics, National Research and Innovation Agency, South Jakarta, Indonesia

*Address all correspondence to: kdsasmita79@yahoo.com

References

[1] ICCO. ICCO Quarterly Bulletin of Cocoa Statistics, 2023;Vol. XLIX, No. 1, Cocoa year 2022/2023. The International Cocoa Organization, Abidjan 06 Côte d'Ivoire

[2] Pusat Data dan Sistem Informasi Pertanian. Outlook Kakao. Sekretariat Jenderal Pertanian: Kementerian Pertanian Republik Indonesia; 2022

[3] Ditjenbun. Statistik Perkebunan Indonesia (2003-2023): Kakao. Direktorat Jenderal Perkebunan, Kementerian Pertanian Republik Indonesia. 2003-2023

[4] Läderach P, Martinez-Valle A, Schroth G, Castro N. Predicting the future climatic suitability for cocoa farming of the world's leading producer countries. Ghana and Côte d'Ivoire. Climatic Change. 2013;**119**:841-854

[5] Gateau-Rey L, Tanner EVJ, Rapidel B, Marelli J-P, Royaert S. Climate change could threaten cocoa production: Effects of 2015-16 El Niño-related drought on cocoa agroforests in Bahia, Brazil. PLoS One. 2018;**13**(7):e0200454. DOI: 10.1371/journal.pone.0200454

[6] Lahive F, Hadley P, Daymond AJ. The physiological responses of cacao to the environment and the implications for climate change resilience: A review. Agronomy for Sustainable Development. 2019;**39**:5. DOI: 10.1007/s13593-018-0552-0

[7] Niether W, Jacobi J, Blaser WJ, Andres C, Armengot L. Cocoa agroforestry systems versus monocultures: A multi-dimensional meta-analysis. Environmental Research Letters. 2020;**15**:104085. DOI: 10.1088/1748-9326/abb053

[8] Weidong W, Yapo C-KLC. Impact of climate change on West Africa cocoa yield: Evidence from cote D'ivoire. International Journal of Economics, Commerce and Management. 2022;**x**(1):163-177

[9] Agbeniyi SO, Adeniyi DO, Ogundeji D. Impact of climate change on timing and frequency of fungicide application for the control of phytophthora pod rot of cacao in Nigeria. In: International Symposium on Cocoa Research (ISCR); 13-17 November 2017; Lima, Peru: ISCR; 2018 pp. unpaginated ref.9

[10] Lima DA, Mizubuti ESG, Andrade GP, de Almeida LA. Cocoa black pod rot incidence as influenced by rainfall and temperature. Phytopathology. 2016;**106**(5):480-487

[11] Cilas C, Bastide P. Challenges to cocoa production in the face of climate change and the spread of pests and diseases. Agronomy. 2020;**10**(9):1232. DOI: 10.3390/agronomy10091232

[12] Ajayi IR, Afolabi MO, Ogunbodede EF, Sunday AG. Modeling rainfall as a constraining factor for cocoa yield in Ondo state. American Journal of Scientific and Industrial Research. 2010;**1**(2):127-134

[13] Chocolate Class. Cacao and Climate Change: Implications and Recommendations. Multimedia Essays on Chocolate, Culture, and the Politics of Food. [Internet]. 2018. Available from: https://chocolateclass.wordpress.com/2018/05/04/cacao-and-climate-change-implications-and-recommendations/

[14] IPCC. Climate change 2007: The physical science basis. In:

Summary for Policymakers. Geneva: Intergovernmental Panel on Climate Change; 2007

[15] Runtunuwu E, Kondoh A. Assessing global climate variability and change under coldest and warmest periods at different latitudinal regions. Indonesian Journal of Agricultural Science. 2008;**9**(1):7-18

[16] KLHK. Roadmap Nationally Determined Contribution (NDC) Adaptasi Perubahan Iklim. Jakarta (ID): Kementerian Lingkungan Hidup dan Kehutanan Republik Indonesia; 2020

[17] Santosa E, Sakti GP, Fattah MZ, Zaman S, Wachjar A. Cocoa production stability in relation to changing rainfall and temperature in East Java. Indonesia. Journal of Tropical Crop Science. 2018;**5**(1):6-17

[18] Ardiani F, Wirianata H, Noviana G. Pengaruh Iklim terhadap produksi kakao di Kabupaten Gunungkidul. Jurnal Agro Industri Perkebunan. 2022;**10**(1):45-52

[19] Towaha J, Wardiana E. Evaluasi tingkat toleransi 35 genotipe kakao terhadap periode kering. Jurnal Tanaman Industri dan Penyegar. 2015;**2**(3):133-142

[20] Sugiatno H, Hansyah AF, Evizal R, Ramadiana S. Pengaruh kemarau terhadap pertumbuhan dan produksi tujuh klon kakao. Jurnal Agrotropika. 2022;**21**(1):59-66

[21] Zakariyya F, Indradewa D, Santoso TI. Changes of leaf anatomical profile of cocoa clones seedlings in response to drought. Pelita Perkebunan. 2019;**35**(3):177-185

[22] Anita-Sari I, Susilo AW, Puspitasari N, Nuraini F, Setyawan B, McMahon P, et al. Intensity of vascular streak dieback in different cocoa clones and various agro-climatic conditions. Pelita Perkebunan. 2017;**33**(1):1-9

[23] Hairiah K, Rahayu S, Suprayogo D, Prayogo C. Perubahan Iklim: Sebab dan dampaknya terhadap kehidupan. Bahan Ajar 1. In: Bogor, Indonesia: World Agroforestry Centre (ICRAF) Southeast Asia Regional Program Dan Malang. Indonesia: Universitas Brawijaya; 2016

[24] Sarjana P, Kasa IW, Gunam IBW, Takama T, Pratiwi L, Putra IWWP. Effects of climate change on the vulnerability of cocoa production in Medewi, Bali, Indonesia. IOP Conf. Series: Earth and Environmental Science. 2021;**724**:012076. DOI: 10.1088/1755-1315/724/1/012076

[25] David M. Kajian ketahanan pada pertumbuhan awal beberapa klon kakao (*Theobroma cacao* L.) terhadap cekaman kekeringan. Surakarta: Tesis Universitas Sebelas Maret; 2008 99 hal

[26] Zakariyya F, Setiyawan B, Susilo AW. Stomatal, proline, and leaf water status characters of some cocoa clones (*Theobroma cacao* L.) on prolonged dry season. Pelita Perkebunan. 2016;**33**(1):109-117

[27] Susilo AW, Anita-Sari I, Setyawan B. Yield performance of promising cocoa clones (*Theobroma cacao* L.) in dry climatic conditions. Pelita Perkebunan. 2020;**36**(1):24-31

[28] Bahrun A, Fahimuddin MY, Rakian TC. Cocoa pod husk biochar reduce watering frequency and increase cocoa seedlings growth. International Journal of Environment, Agriculture and Biotechnology. 2018;**3**(5):1635-1639

[29] Tee YK, Abdul Haadi H, Raja Norsharizatul Azira RMS. Stress tolerance of cacao trees (*Theobroma cacao* L.) subjected to smart water gel®.

Malaysian Society of Plant Physiology Conference Trans. Malaysian Soc. Plant Physiology. 2018;**25**:2600-9595

[30] Mintah LO, Ofosu-Budu GK, Osei-Bonsu NO, Ulzen J, Danso EO. Growing media, water stress and re-watering effects on the growth and dry matter production of cocoa seedlings. Acta Universitatis Sapientiae, Agriculture and Environment. 2022;**14**(1):45-61

[31] Sasmita KD. Aplikasi arang, pupuk organik, dan mikrob pelarut fosfat untuk perbaikan sifat tanah masam dan peningkatan keefektifan pupuk P pada bibit kakao. Disertasi IPB Bogor; 2017

[32] Prayudyaningsih R. Mikoriza dalam Pengelolaan Hama-Penyakit Terpadu di Persemaian. Info Teknis Eboni Balai Penelitian Kehutanan Makassar. 2012;**9**(1):55-75

[33] Lucía A, Pardo J, Durántez A, Hoyos J, Chicharro JL. Physiological differences between professional and elite road cyclists. International Journal of Sports Medicine. 1998;**19**(5):342-348

[34] Kaba JS, Abunyewaa AA, Kugbeb J, Kwashiea GKS, Ansaha EO, Andoha H. Arbuscular mycorrhizal fungi and potassium fertilizer as plant biostimulants and alternative research for enhancing plants adaptation to drought stress: Opportunities for enhancing drought tolerance in cocoa (*Theobroma cacao* L.). Sustainable Environment. 2021;**7**(1):1-12

[35] Droh G, Kouassi A-B, Kouadjo ZGC, Zeze A, Nguetta AS-P, Sanders IR. Effects of two types of AMF on growth of cocoa seedlings (*Theobroma cacao* L.) in greenhouses. Global. Journal of Advanced Research. 2016;**3**(3):157-164

[36] Ramírez JG, Osorno L, Osorio NW. Presence of mycorrhizal fungi and a fluorescent *pseudomonas sp*. in the rhizosphere of cacao in two agroecosystems and their effects on cacao seedling growth. Agronomía Colombiana. 2016;**34**(3):385-392

[37] Famuwagun IB, Agele SO. Cacao Growth and Development under Different Nursery and Field Conditions. In: Aikpokpodion PO, editor. Theobroma Cacao – Deploying Science for Sustainability of Global Cocoa Economy. London: IntechOpen. 2019. DOI:10.5772/intechopen.73761. ISBN:978-1-83962-733-0.

[38] Muzlifa R, Fikrinda, Jufri Y. Pengaruh fungi mikoriza arbuskula dan kompos limbah kakao terhadap kolonisasi mikoriza, dan pertumbuhan bibit kakao pada ultisol. Jurnal Ilmiah Mahasiswa Pertanian. 2019;**4**(4):657-665

[39] Pattirane MCPA, Mahulette AS, Marasabessy DA. Respon pertumbuhan bibit kakao (*Theobroma cacao* L.) pada berbagai dosis pemberian rock phosphate dan fungi mikoriza arbuskular (FMA). Jurnal Agrohut. 2022;**13**(2):77-86

[40] Mowidu I, Endang SDHS, H.S. Rekayasa lingkungan perakaran melalui sistem rorak untuk meningkatkan serapan hara pada tanaman kakao. Agropet. 2022;**19**(1):9-15

[41] Beer J, Muschler R, Kass D, Somarriba E. Shade management in coffee and cacao plantations. Agroforestry Systems. 1998;**38**:139-164

[42] Moser G, Leuschner C, Hertel D, Holscher D, Kohler M, Leitner D, et al. Response of cocoa trees (*Theobroma cacao*) to a 13-month desiccation period in Sulawesi, Indonesia. Agroforestry Systems. 2010;**79**:171-187. DOI: 10.1007/s10457-010-9303-1

[43] Schwendenmann L, Veldkamp E, Moserz G, Holscher D, Kohler M,

Clough Y, et al. Effects of an experimental drought on the functioning of a cacao agroforestry system, Sulawesi, Indonesia. Global Change Biology. 2010;**16**:1515-1530. DOI: 10.1111/j.1365-2486.2009.02034.x

[44] Dhanapal R. Relevance and opportunities in coconut based cropping/farming systems. In: Thomas GV, Krishanakumar V, Maheswarappa HP, Palaniswami C, editors. Coconut Based Cropping/Farming Systems. Kasaragod: Central Plantation Crops Research Institute; 2010. pp. 1, 231-8

[45] Maheswarappa HP, Dhanapal R, Subramanian P, Palaniswami C. Evaluation of coconut based high density multi-species cropping system under organic and integrated nutrient management. Journal of Plantation Crops. 2013;**41**(2):130-135

[46] Krishnakumar V, Maheswarappa HP. Coconut based cropping systems in root (wilt) diseased areas. In: Thomas GV, Krishanakumar V, Maheswarappa HP, Palaniswami C, editors. Coconut Based Cropping/Farming Systems. Kasaragod: Central Plantation Crops Research Institute; 2010. pp. 62, 231-79

[47] Nurmi, Haridjaja O, Arsyad S, Yahya S. Perubahan sifat fisik tanah sebagai respon perlakuan konservasi vegetatif pada pertanaman kakao. Forum Pascasarjana. 2009;**32**(1):21-31

[48] Prawoto AA, Sholeh PNM. Produksi awal dan kajian ekonomis usahatani nilam Aceh (*Pogostemon cablin* Benth.) sebagai tanaman sela kakao muda. Pelita Perkebunan. 2006;**22**(3):168-190

Section 2

ICT Role in Advancing Climate Change Adaptation and Climate Mitigation

Chapter 6

Using Information Technologies (ICTs) to Improve *Theobroma cocoa* Extension Service: Lessons from the Case of Ghanaian Bean Farmers

Julia Bello-Bravo and Anne Namatsi Lutomia

Abstract

Ghana's modern cocoa production depends on farmers having access to innovations and updated best practices delivered through "new" information and communication technology (ICT) channels. However, extension services continue to face familiar delivery shortfalls affecting both the national-level extent of Ghana's cocoa production and the local livelihoods of its producers. This chapter draws on questionnaire data collected during a training workshop for postharvest loss prevention with mostly university-educated farmers to explore how they access innovation and best practices information through new and old technological channels. Key findings indicate that while farmers utilize both old and new ICTs, they still experience the familiar barriers of low agricultural extension agent-to-farmer ratios and shortages of resources. Recommendations include (1) ensuring that the affordances of "old" technologies are amplified rather than replaced by "new" ones, and (2) the use of highly scalable educational animations deployed individually for training or through virtual communities of practices to "bypass" the perennial issue of resource shortfalls in extension services.

Keywords: video animations, cocoa, farmers, ICTs, Ghana

1. Introduction

Theobroma cacao, commonly known as cocoa or cacao, is a tropical tree species belonging to the Malvaceae family. Native to the Amazon basin and other regions of South and Central America, the term "Theobroma" is derived from the Greek words "theo" ("deity") and "broma" ("food"), hence the notion that cocoa is the "food of gods." *Theobroma cacao* is cultivated primarily for its seeds, used to produce cocoa powder, cocoa butter, and chocolate products [1]. Presently, cocoa is widely cultivated in many parts of the world, including West Africa, where it is a significant source of income for many smallholder farmers.

Ghana is internationally known for its cocoa production and trade. As the second-leading world cocoa producer, it has registered an annual minimum cocoa bean output of 700,000 metric tons since 2012. For instance, the 2020/2021 crop season recorded an estimated 1.1 million metric tons in cocoa beans production. The Ghana Cocoa Board (COCOBOD), the government institution responsible for regulating cocoa buying prices, identified some threats to cocoa production in Ghana [2].

Research on smallholder farmers in Africa, including Ghana, has long indicated significant time and cost savings from using information and communication technology (ICT) for extension services [3]. Agriculture information benefits smallholder farmers by improving food production and increasing agriculture development and information on farming practices, market prices, and disease and pest control. The proliferation of mobile phones has been critical for smallholder farmers because they can increase agro-entrepreneurship by allowing efficiency in accessing markets, promoting investment, and empowering them [4]. However, Ghana's cocoa production faces challenges regarding extension service delivery to cocoa farmers. Inadequate extension services, limited access to information, poor infrastructure, language barriers, and limited funding are key challenges that cocoa farmers face [2]. Both old and new technologies are used to access this information. Old technologies include communication tools and modes of communication prior to the advent of computers and mobile phones, e.g., typewriters, landlines, VCRs. New technologies include smartphones, tablets, computers, artificial intelligence, and cloud computing that can center use experience.

The proliferation of affordable ICTs has paved the way for knowledge transfer in the agricultural community around the world using local languages and video-capable devices [5, 6]. Another benefit of mobile phones is speeding up access to, adapting, and sharing information [7]. Relatedly, Overå [8] found that farmers join social and business networks to seek useful and relevant information, which helps them to make better decisions for profitable, efficient, and productive agricultural production [9].

This chapter aims to identify how cocoa farmers in Ghana access information for better cocoa production using old and new technologies. It asks: (1) What kind of devices do farmers own? (2) What sources of information do farmers access for relevant information about beans? (3) What types of information do these sources provide? (4) How do farmers rate this information? and (5) How can the farmers learn to use their devices to access video animations? The chapter draws on the experiences of 15 male cocoa farmers' responses to a survey at a workshop for using a hermetic jerrycan storage technique to prevent postharvest bean loss [10]. Our findings generate recommendations for the use of old and new technology pathways in disseminating agricultural information to improve cocoa production.

2. Scientific animations without Borders and agriculture education

Founded in 2011, Scientific Animations Without Borders (SAWBO) creates educational animations in local languages shareable on ICTs (especially cell phones) for adult learners with little or no print-literacy [11]. These educational animations cover over 140 topic areas in more than 280 languages and dialects—especially around the topics of agriculture, health, women's empowerment, and peace-building—and have reached 50+ million people since the program's inception. These videos can be downloaded free of charge and shared within communities [11, 12]. SAWBO's primary goal has been a systems approach to knowledge chains that generate educational materials on best practices accessible to and adopted by the widest possible demographic [13, 14].

A knowledge chain consists of four process links and the transitions between them [15]: (1) to identify and scientifically frame a problem so that solutions to that framing of the problem are implied, (2) to embody that scientifically abstract solution to the problem in a concrete, animated video—overdubbed as needed into any locally accurate dialect—as a means of effecting knowledge transfer about the importance, context, and need to act on the solution offered for the problem in the video, (3) to disseminate and maintain an infrastructure that affords not only free and easy access to SAWBO videos for educators, learners, and researches but also a means for SAWBO to collect and incorporate feedback on any video or part of the knowledge chain itself, and (4) to look for ways to more widely scale the reach of any video and the processes that developed it. This chain includes loop-backs that support later modification of videos produced, above all translation into new languages and dialects that affords dramatically scaling-up the reach of any video at decreasing unit cost [14]. This approach has allowed individuals and organizations of any size, from small community-based organizations to large government and international organizations, to access, select, and adapt scientifically accurate educational content in local languages relevant to local problems in need of interventions [15].

3. Extension service and cocoa production

Sustaining the interest of cocoa farmers in the sector requires significant investment and research into the welfare and sustainability of their farming practices. The majority of cocoa production in Ghana is carried out by smallholders who cultivate on land plots <2 hectares using traditional production methods, resulting in low yields [16]. To enhance agricultural productivity and alleviate poverty in these areas, disseminating modern agricultural technologies to rural farmers is crucial [17].

Unfortunately, the growth of rural cocoa-growing communities and the agricultural sector is hindered by inadequate and ineffective delivery of extension services [18]. Given the high global demand for cocoa beans and the low productivity levels of approximately 90% of Ghanaian farms, extension services are essential for developing the human capacity of farmers and providing them with financial support [18].

In the twenty-first century, the role of extension services has in principle, expanded to encompass more than simply technology transfer and production improvement [19]. Agricultural extension programs aim more holistically to facilitate systematic development and improvement of the farming environment, reduce poverty, enhance agricultural production, promote processing capabilities, and teach large groups of farmers new knowledge and best practices, ultimately generating improved productivity and income [19].

3.1 Digital divide and cocoa production in Ghana

The gender gap and smallholder production are impacted by Ghana's digital divide in the cocoa industry. Digital literacy, internet connectivity, and access to digital technologies are significant drivers of the divide, which impacts many facets of cocoa production. Limited access to digital technologies and the Internet prevents smallholder farmers (male and female) from accessing vital information and knowledge resources [20]. These barriers and lack of access impede productivity and hinder farmers' capacity to adapt to climate change and other challenges. The digital divide also excludes women and smallholder farmers from accessing financial services, thus increasing socio-economic gender disparities [21]. Another impact of the digital divide is the lack of access to markets and value chain integration. Digital platforms and e-commerce provide farmers access to markets, fair prices, and direct contact with buyers [22].

Importantly, while low or no print-literacy affects access to digital information [23], it does not necessarily impact the capacity to learn through digital means [13]. Equally, while higher levels of education often associate with increased use of digital information [22], this does not guarantee farmers' ability to access such information. Lack of access to the Internet and digital technologies means farmers cannot connect to markets and gain market insights. Consequently, their capacity to interact with customers, learn about the market, and engage in fair competition along the cocoa value chain is constrained. A lack of training and unfamiliarity with digital tools can further hinder the ability to navigate online platforms, access information, and leverage digital resources for improved farming practices [22]. Bridging the digital literacy gap through targeted training programs can empower smallholder farmers to make more informed decisions and enhance productivity. Addressing the digital divide in cocoa production is critical to giving smallholder farmers' voices more power; closing gaps will afford more power over farming operations and decision-making and access to financial services [20].

Improving internet connectivity, developing rural infrastructures, providing inexpensive access to digital technologies, and providing focused digital literacy initiatives for smallholder farmers (including women) should be priorities. To guarantee that the advantages of the digital revolution in cocoa production are available to all stakeholders, public–private partnerships, legislative efforts, and interventions should promote inclusive and sustainable strategies.

4. Case study of Ghanaian cocoa farmers

4.1 Method

Data for this study is drawn from a training workshop in Ghana in 2017 that provided improved agricultural technological tools and skills to farmers using a 4.5-minute-long animated SAWBO video. While the immediate goal of the workshop focused on learning gains around the improved postharvest loss prevention technique using jerrycans and willingness to adopt it as a practice, data collected from 15 male farmers also included modes of access they utilized for obtaining agricultural information. A survey questionnaire was administered in Fante (the locally most comfortably spoken language among participants) to measure learning gains, willingness to adopt the loss prevention technique, qualitative reactions to the video itself, and pathways to obtaining agricultural information.

Participants were 15 male cocoa farmers by profession from the central region of Ghana, purposefully sampled from five villages for the postharvest loss prevention training. Quantitative responses were tabulated and analyzed in Excel (**Table 1**).

5. Findings

Device ownership affects extension service information access. Farmers in the study had both "old" (TV, radio) and "new" technologies (computers, tablets, and

Characteristic	Response Categories	Number	Frequency %
Gender	Male	15	100
Marital status	Divorced	1	7
	Single	1	7
	Married	13	87
Education status	Tertiary education (university degree	13	87
	High school and diploma)	2	13
Group of age in years	18–29	2	13
	30–50	9	60
	51–70	4	27

Table 1.
Demographic data of the responders.

mobile phones, with and without video capability). All participants possessed the "old" access modes, with the majority (86%) owning smartphones. Similarly, although the majority owned PC computers, the majority also did not own laptops. (See **Table 2**).

The participants reported their perceptions about the video animation. Most heard the training clearly, and ten reported the visual quality was mostly clear. Regarding the strengths of the video for understanding, 15 participants reported that

Device Ownership	Response	Number	Frequency %
Radio	No	0	0
	Yes	15	100
TV	No	0	0
	Yes	15	100
Desktop Computer	No	3	20
	Yes	12	80
Laptop	No	13	87
	Yes	2	13
Tablet	No	6	40
	Yes	9	60
Smartphone	No	1	6
	Yes	13	86
Video-capable cellphone	No	0	0
	Yes	1	6
Nonvideo-capable cellphone	No	0	0
	Yes	0	0

Table 2.
Device ownership.

the video's language was totally understandable. All reported the picture quality as good and demonstrated the technique. The participants found no weaknesses in the videos and provided no comments for improvement (see **Table 3**).

Regarding the usefulness of the video for training, learning, and technical accuracy, all participants indicated the videos were very useful as learning resources and offered completely technically correct information. They also indicated that they found the video animation presentation very interesting (see **Table 4**).

Farmers answered "yes" or "no" to the relevance of various sources of information for cocoa production. In general, all information pathways were noted as relevant. Two participants replied that politicians were not relevant sources of information (See **Table 5**).

How farmers rate the perceived or actual usefulness of information from various sources is important. All information pathways were noted as very useful, with the exception of two participants who identified government extension staff and politicians as somewhat useful (see **Table 6**).

Farmers reported that they received relevant information from varying sources of information. Radio, smartphones, video-capable phones, neighbors/friends, newspapers, and government extension services provided similar information about bean

Questions	Responses	Number	Frequency %
Could you hear the training clearly	Mostly clear	2	13
	Totally clear	13	86
Was the visual quality good?	Mostly clear	10	67
	Totally clear	5	23
Strengths of the video for understanding	Action and demonstration	15	100
	High-quality pictures	15	100
Strengths of the video for understanding	Language understandable	15	100
	High-quality pictures	15	100
Weakness of the video for understanding	The videos had low sound	0	0
	The video was too short	0	0
	Pictures were not clear	0	0
	The video was too fast	0	0
	The video was slow at times	0	0
What suggestions do you have for improving the video used in training?	The video should be slow	0	0
	The sound should be louder	0	0
	The container was not transparent	0	0
	The video should be low	0	0
	The picture should be clear	0	0
	The training should be done in communities and farms	15	100

Table 3.
Perception of video animations and suggestions for improvement.

Usefulness of videos	Usefulness of videos	Number	Frequency %
Usefulness in training	Very useful	15	100%
The usefulness of video in learning	Very useful	15	100%
Technical correctness	Very useful	15	100%

Table 4.
Usefulness of video in training, learning, and technical correctness.

Sources	Relevant to bean production			
	Yes	Frequency %	No	Frequency %
Radio	15	100	0	0
Television	15	100	0	0
Smartphone	15	100	0	0
Video-capable cellphone	1	6	0	0
Neighbors/Friends	15	100	0	0
Family members	15	100	0	0
Newspapers	15	100	3	20
NGO Extension staff	15	100	0	0
Government Extension staff	13	87	2	13
Politicians	13	87	2	13

Table 5.
Sources of information relevant to bean production.

Sources	Useful	Number	Frequency %
Radio	Not very useful	0	0
	Somewhat useful	0	0
	Very useful	15	100
	Not very useful	0	0
Television	Somewhat useful	0	0
	Very useful	15	100
Smartphone	Not very useful	0	0
	Somewhat useful	0	0
	Very useful	15	100
Video-capable cellphone	Not very useful	0	0
	Somewhat useful	0	0
	Very useful	1	6
Neighbors/Friends	Not very useful	0	0
	Somewhat useful	0	0
	Very useful	15	100

Sources	Useful	Number	Frequency %
Family members	Not very useful	0	0
	Somewhat useful	0	0
	Very useful	15	100
Newspapers	Not very useful	0	0
	Somewhat useful	0	0
	Very useful	15	100
	Not very useful	0	0
NGO Extension staff	Somewhat useful	0	0
	Very useful	15	100
Government Extension staff	Not very useful	0	0
	Somewhat useful	2	13
	Very useful	13	87
Politicians	Not very useful	0	0
	Somewhat useful	2	13
	Very useful	13	87

Table 6.
Usefulness of information relevant to bean production.

farming seeds, planting, growing and storage, and pricing beans. Television provided the most information and showed farming demonstrations. The NGO extension staff focuses on postharvest loss prevention. In contrast, government extension service staff and politicians uniquely focus on bean production and seed and bean prices for the market. Politicians provided information about producing beans, seeds, and pricing beans (**Table** 7).

Sources	Relevant information
Radio	Bean farming-seeds, planting, growing and storage, pricing beans
Smartphone	Bean farming-seeds, planting, growing and storage, pricing beans
Video-capable phone	Bean farming-seeds, planting, growing and storage, pricing beans
Television	Showing demonstrations and farmers' experience about bean production: seeds, planting, growing, postharvest loss prevention, and storage, pricing beans
Neighbors/Friends	Bean farming-seeds, planting, growing and storage, pricing beans
Family members	Bean farming-seeds, planting, growing and storage, pricing beans
Newspapers	Bean farming-seeds, planting, growing and storage, pricing beans
NGO Extension staff	Postharvest loss prevention
Government Extension staff	Bean farming-seeds, planting, growing and storage, pricing beans
Politicians	Producing beans, seeds, and pricing beans

Table 7.
Sources and categories of relevant information.

Farmers were asked to rate the quality of categories of information on an 8-point scale (from 0 to 7, low to high). All sources of information were rated 7, with the exception of government extension staff and politicians, which two participants rated as 6 (**Table 8**).

When asked if they intended to share the postharvest loss prevention video animation with others after the workshop, all the participants said they would certainly share it (**Table 9**).

The farmers were asked whether people could download and upload the animation on a desktop, laptop, smartphone, or video-capable phone. All of the participants indicated this was possible (see **Table 10**).

Sources	Ranking								Total responses
	0	1	2	3	4	5	6	7	
Radio								15	15
Smartphone								15	15
Video-capable phone								15	15
Television								15	15
Neighbors/Friends								15	15
Family members								15	15
Newspapers								15	15
NGO Extension staff								15	15
Government Extension staff							2	13	15
Politicians							2	13	15

Table 8.
Ranking of quality of information by sources.

Activity	Sharing certainty	Number	Frequency%
Sharing video	Certainly not share	0	0
	Not sure whether to share	0	0
	Probably share	0	0
	Certainly share	15	15

Table 9.
Intention to share the video with others after the training.

Ability to download and upload animation to devices	Number	Frequency %
Desktop computer	15	100
Laptop	15	100
Smartphone	15	100
Video-capable phone	15	100

Table 10.
Ability to download and upload animations on various devices after practice.

Confidence level in downloading and uploading videos to devices	Not confident at all	Not very confident	Somewhat confident	Very confident	Not sure
Desktop computer	0	0	0	15	0
Laptop	0	0	0	15	0
Smartphone	0	0	0	15	0
Video-capable phone	0	0	0	15	0

Table 11.
Confidence level in uploading and downloading video animations to devices.

The farmers were also asked to indicate whether people could confidently upload and download video animations on devices such as desktop computers, laptops, smartphones, and video-capable phones confidently. All participants indicated that they were very confident this could be done (**Table 11**).

6. Discussion

Across Africa, many smallholder farmers are accessing information, services, and products to boost crop yields and increase efficiency and incomes using digital technologies. Smallholder farmers currently benefit from various innovations in agriculture or agri-tech. These innovations include mobile apps for inputs such as seeds and weather alerts. There are also other farming precision solutions such as satellite, drone imagery, and sensors that provide real-time data about the health of crops.

The farmers reported using mobile phones to communicate with other farmers and extension service officers. They also added that they use mobile phones to access information more quickly. Consistent with other research [24–26], challenges these farmers face associated with using mobile phones include network failure, high cost of data bundles, and lack of reliable electricity for charging phones.

Although old forms of accessing information—including radio, word of mouth, home visits, and extension services—persist, these are not necessarily efficient, adaptive, or fast enough on their own and benefit from working hand-in-hand with new ICTs and mobile phones [27, 28].

Informal agricultural knowledge is passed down to family members, neighbors, or both. Farmers may easily exchange information with other farmers and between generations, which requires no considerable transaction costs [19]. Our research supports the notion that farmers still rely on word of mouth from their friends, neighbors, and families, but such access now also occurs via mobile phones.

In contrast, formal information channels offer (1) precise guidance on workable adaption strategies, (2) information in understandable terms that focus on cocoa smallholders, and (3) information responsive to pressing livelihood concerns. Our research demonstrates that farmers evaluate the information they receive as relevant or not; while this means that information must be tailored to meet the farmers' needs [29], it is unfortunate that politicians and government extension officers could be perceived as less relevant. Reasons for this were not given by the two participants who noted this; however, research also indicates that a combination of animated viewing and facilitated discussion afterward achieves higher learning gains than traditional extension teaching in isolation [13]. This may be a way to overcome a perceived relevance shortfall in extension service providers.

Cocoa production in Ghana is not culturally neutral. Concerning the cocoa industry, politicians play a significant role in shaping and informing policies, programs, and resource allocation, which impact farmers and the cocoa sector [30]. Unlike other information pathways, which had a 100% consensus on the relevance and usefulness of the information, a majority of farmers indicated that information received from politicians was also useful and relevant, but two did not find it optimally so.

Lastly, all of the participants said they would "certainly share" the video animation with others and were confident of their ability to access and download the video for such use in the future. Participants appreciated the possibility of re-watching part or all of an animated video as a refresher and the ability to play, stop, fast-forward, and replay the videos, consistent with other research; for example, Van Mele [31] highlighted how this kind of flexibility in video media facilitates effective training and learning in developing countries.

7. Recommendations

Notwithstanding the enthusiasm for new technologies [32], existing infrastructures of "old" technologies still afford communicating important agricultural information to farmers. Specifically, they continue to be very well-suited for delivering educational and best-practices information, including educational materials on SAWBO's videos. The main inefficiencies of these channels, which should not be abandoned, include less ability to adaptably, flexibly, or in real-time deliver newly emerging information to farmers, especially about markets. The appeal, usability, and technological familiarity of mobile phones for providing such flexible, adaptive, and real-time information make them exceptionally promising [33], providing that they do not reproduce or worsen existing social inequalities, particularly around access by female farmers.

Although a recommendation to strengthen extension service delivery by increasing national investment in it faces the perennial shortfalls of agent and budgetary allocations, the recommendation must still be made. Here again, the use of animated video helps increase extension's capacity by reducing its budgetary overhead [14]; these can be (and are being) delivered via "old" technologies. Extension effects can also be amplified and supported by using social media (WhatsApp) groups to form virtual communities of practice for supporting agricultural innovations and the adoption of best practices [15]. Such communities of practice (both virtual and offline) can generate context-specific solutions to local problems, the testing, development, or adaptation of technologies [34], and thus the dissemination of formal and citizen-generated research to farmers. For food security, as perhaps the most important backbone of the Sustainable Development Goals, extension services can link up with grassroots, citizen-led, or community-of-practice activities to train farmers on postharvest loss prevention, increased quality standards to enhance crop yields, and other value-addition techniques. These increase the profitability of cocoa farming and improve the livelihoods, well-being, and quality of life for farmers. Lastly, further work must be done to identify more channels for reaching the widest demographic. In the face of the changing climate and the future of the globe generally, we can no longer afford to miss people with life-critical information and skills.

8. Limitations and future research

The major limitation on the generalizability of this study is its small sample size of male, predominantly university-level educated participants. However, SAWBO has designed and tested its video media for effectiveness with the widest demographic possible, regardless of age, gender, educational and technological literacy, or geographic (rural/urban) isolation [13, 35]. The findings of this study for appeal, willingness to share, and video knowledge transfer for this particular demographic agree with previous research [10]. The limited sample also reflects the technological familiarity of more highly educated groups; that is, while mobile phones are globally the most common digital access device, laptops, and tablets are much less common for people in Africa [5]. The high prevalence of PC computers is consistent with exposure to computers among university-educated groups [36]. This highlights that future research to investigate modes of old and new information access by women is necessary, especially as women typically experience more limited access to education and digital technologies often associated with them [36–38].

9. Conclusions

The diffusion of "new" technologies always occurs through the infrastructures and channels of existing technologies, now becoming "old." In this study, while the predominantly university-educated participants had the new digital access technologies of PC computers and mobile phones, they also all had televisions and radios. When focusing on the promising affordances of new technologies, we should not overlook or even seek to supplant the old technologies. For example, these days books are in many ways obsolete, and yet they still have certain affordances, advantages, pleasures, and cultural habits that would disappear if completely replaced by digital "texts."

This point has direct bearing on attempts to innovate within a still-predominantly agricultural society like Ghana's cocoa production. We are currently in the International Decade of Indigenous Languages, as part of a global effort to preserve and honor indigenous knowledge, thinking, and practices. This is not solely to "archive" such knowledge, thinking, and practices like fossils or museum pieces in the event of their extinction but also to draw upon the various wisdoms contained within them for the benefit of humanity general as it faces the prospect of its extinction outright. The recommendation to not supplant or entirely replace Ghana's "old" technologies for agricultural support for cocoa with "new" technologies is not merely strategic (because the transition would then be smoother). It is also because traditional agricultural settings are themselves rooted in wisdom of adapting, amplifying, and modifying old practices, usually in sustainably "slower" ways.

Ghanaian farmers rely on old and new ICTs to access information, whether radios, television, smartphones, or word of mouth. Information access barriers include a lack of appropriate extension service, low agricultural extension agent-to-farmer ratios, and a digital divide in accessing ICTs, especially by women. As always, smallholder farmers need (1) access to relevant information to improve cocoa productivity, (2) in relatable and usable language and forms, and (3) both formal and informal access to ICT-amplified extension services [39]. Maintaining a continuity between the "old" and the "new," and ensuring the fewest disruptions without reproducing or worsening existing social inequalities, is essential to this.

Acknowledgements

This research comprises a QED Group LLC and USAID-funded project. The funding aimed to generate 'how-to' videos for agricultural techniques or technologies. Dr. Samuel Amoa-Mensa, director of the Center for Learning and Community Engagement (CLCD) in Cape Coast, Ghana, facilitated the data collection during the study. Many thanks to our funders and Dr. Amoa-Mensa for being the host institution during our fieldwork.

Conflict of interest

The authors declare no conflict of interest.

Author details

Julia Bello-Bravo and Anne Namatsi Lutomia*
Purdue University, West Lafayette, USA

*Address all correspondence to: alutomia@purdue.edu

References

[1] Verna R. The history and science of chocolate. The Malaysian Journal of Pathology. 2013;**35**(2):111-121

[2] Ghana Cocoa Board. Ghana Cocoa Board. 2023. Available from: https://cocobod.gh/

[3] Aker JC. Dial "a" for Agriculture: Using Information and Communication Technologies for Agricultural Extension in Developing Countries (Working Paper 269). Washington, DC: Center for Global Development; 2010

[4] Folitse BY et al. The determinants of mobile-phone usage among small-scale poultry farmers in Ghana. Information Development. 2019;**35**(4):564-574

[5] Bello-Bravo J et al. Breaking out: The turning point in learning using mobile technology. Heliyon. 2021;**7**(3):e06595

[6] FAO. Food and Agriculture: Driving Action across the 2030 Agenda for Sustainable Development. Rome, IT: FAO; 2017

[7] Nyamba SY, Mlozi MR. Factors influencing the use of mobile phones in communicating agricultural information: A case of Kilolo District, Iringa, Tanzania. International Journal of Information and Communication Technology Research. 2012;**2**(7):558-563

[8] Overå R. Networks, distance, and trust: Telecommunications development and changing trading practices in Ghana. World Development. 2006;**34**(7):1301-1315

[9] Lwoga ET. Bridging the agricultural knowledge and information divide: The case of selected telecenters and rural radio in Tanzania. The Electronic Journal of Information Systems in Developing Countries. 2010;**43**(1):1-14

[10] Bello-Bravo J et al. An 89% solution adoption rate at a two-year follow-up: Evaluating the effectiveness of an animated agricultural video approach. Information Technology for Development. 2020;**26**(3):577-590

[11] Rodríguez-Domenech MÁ, Bello-Bravo J, Pittendrigh BR. Scientific animations without Borders (SAWBO): An innovative strategy for promoting education for sustainable development. Sustainability Science. 2019;**14**(4):1105-1116

[12] Maredia MK et al. Can mobile phone-based animated videos induce learning and technology adoption among lowliterate farmers? A field experiment in Burkina Faso. Information Technology for Development. 2018;**24**(3):1-32

[13] Bello-Bravo J et al. Facilitated discussions increase learning gains from dialectically localized animated educational videos in Niger. Information Technology for Development. 2018;**24**(2):1-25

[14] Bello-Bravo J et al. Dramatically increased accessibility and decreased cost-per-person impacts are needed for scaling IPM in Africa. Current Opinion in Insect Science. 2022;**54**:100971

[15] Bello-Bravo J et al. Gender, Digitalization, and Resilience: Failing Forward. New York City, NY: Routledge; 2023

[16] Danso-Abbeam G et al. Technical efficiency in Ghana's cocoa industry: Evidence from Bibiani-Anhwiaso-Bekwai District. Journal of Development

and Agricultural Economics. 2012;**4**(10):287-294

[17] Asfaw S et al. Impact of modern agricultural technologies on smallholder welfare: Evidence from Tanzania and Ethiopia. Food Policy. 2012;**37**(3):283-295

[18] Binam JN et al. Factors affecting the technical efficiency among smallholder farmers in the slash and burn agriculture zone of Cameroon. Food Policy. 2004;**29**(5):531-545

[19] Feder G, Savastano S. Modern agricultural technology adoption in sub-Saharan Africa: A four-country analysis. In: Pingali P, Feder G, editors. Agriculture and Rural Development in a Globalizing World: Challenges and Opportunities. New York City, NY: Routledge; 2017. pp. 11-25

[20] Ohemeng FLK, Ofosu-Adarkwa K. Overcoming the digital divide in developing countries: An examination of Ghana's strategies to promote universal access to information communication technologies (ICTs). Journal of Developing Societies. 2014;**30**(3):297-322

[21] Abdulai A-R. The Digitalization of Agriculture and the (Un) Changing Dynamics of Rural Smallholder Farming Systems in Ghana, Sub-Sahara Africa. Guelph, Ontario, CN: University of Guelph; 2022

[22] Abdulai A-R, Bahadur K, Fraser E. What factors influence the likelihood of rural farmer participation in digital agricultural services? Experience from smallholder digitalization in northern Ghana. Outlook on Agriculture. 2023;**52**(1):57-66

[23] Brown A. Participatory Digital Extension in Agriculture: A Study from Ghana. Guelph, Ontario, CN: University of Guelph; 2023

[24] Bello-Bravo J et al. Making agricultural learning accessible: Examining gender in the use of animations via mobile phones. In: Mills M, Wake D, editors. Empowering Learners with Mobile Open-Access Learning Initiatives. Hershey, PA: IGI Global; 2017. pp. 74-100

[25] Gupta A, Ponticelli J, Tesei A. Information, Technology Adoption and Productivity: The Role of Mobile Phones in Agriculture. New York City, NY: National Bureau of Economic Research; 2020

[26] Porter G et al. Mobile phones, gender, and female empowerment in sub-Saharan Africa: Studies with African youth. Information Technology for Development. 2020;**26**(1):180-193

[27] Gupta A, Ponticelli J, Tesei A. Information, Technology Adoption and Productivity: The Role of Mobile Phones in Agriculture. Vol. No. w27192. Cambridge, MA: National Bureau of Economic Research; 2022

[28] Bello-Bravo J et al. Making agricultural learning accessible: Examining gender in the use of animations via mobile phones. In: Khosrow-Pour M et al., editors. Environmental and Agricultural Informatics: Concepts, Methodologies, Tools, and Applications. Hersey, PA: IGI; 2020. pp. 716-736

[29] Knowles MS, Holton EF, Swanson RA, editors. The Adult Learner. 7th ed. New York City, NY: Routledge; 2012

[30] Attipoe SG et al. Assessing the impact of non-governmental organization's extension programs on sustainable cocoa production and household income in Ghana. Journal of Integrative Agriculture. 2021;**20**(10):2820-2836

[31] Van Mele P. Video-Mediated Farmer-to-Farmer Learning for Sustainable Agriculture. A Scoping Study for SDC, SAI Platform and GFRAS. Ghent, Belgium: Agro-Insight; 2011

[32] Bello-Bravo J. "New wine in old bottles": Structures of feeling for a new way to see new wood products. Forests. 2023;**14**(3):524

[33] Aker JC, Mbiti IM. Mobile phones and economic development in Africa. The Journal of Economic Perspectives. 2010;**24**(3):207-232

[34] Bello-Bravo J et al. Identifying and evaluating farmer deviations from steps recommended for hermetic postharvest storage of beans in northern Mozambique. Journal of Stored Products Research. 2020;**87**:101628

[35] Bello-Bravo J, Rodríguez- Domenech MÁ, Pittendrigh BR. SDG1 in Europe: Micro-Grants, Poverty, and the Big-Picture Future of Sustainable Development in a Post-Pandemic World, in Sustainable Development Goals in Europe: A Geographical Approach, M.L. De Lázaro Torres and R. De Miguel González, Editors. Cham. CH: Springer; 2023. p. 3-20

[36] Williams CD, Pitchforth EL, O'Callaghan C. Computers, the internet and medical education in Africa. Medical Education. 2010;**44**(5):485-488

[37] Mariscal J et al. Bridging the gender digital gap. Economics. 2019;**13**(1):20190009

[38] Mossberger K, Tolbert CJ, McNeal RS. Digital Citizenship: The Internet, Society, and Participation. Cambridge, MA: MIT Press; 2008. p. 248

[39] Ilboudo J-P, del Castello R. Linking rural radio to new ICTs in Africa: Bridging the rural digital divide. In: Girard B, editor. The One to Watch Radio, New ICTs and Interactivity. Rome, IT: FAO; 2003. pp. 45-61

Section 3

Climate Change, Crop Protection and the Environment

Chapter 7

Evaluation of Progress in Cocoa Crop Protection and Management

Alex Asante Appiah

Abstract

Cocoa cultivation began with the Olmecs, who were the first humans to consume chocolate as a drink in equatorial Mexico between 1500 and 400 BC. Over the centuries, commercial cocoa cultivation and trade have developed from the Mayans, Aztecs, and through Meso-America under the influence of the Spanish explorers. In 1822, cocoa was first introduced to São Tomé and Príncipe in Africa from where it spread as a plantation crop, with West Africa becoming the major centre of global production. The cultivation of selected hybrid varieties particularly have led to pest and diseases becoming major production limiting factors. This chapter evaluates crop protection techniques developed over the years, and highlights their contribution to yields, production costs, impact on farmers, and the cocoa value chain and ecosystems. We discussed the need to re-evaluate the imbalance of power in the global value chain, the colonial trading systems, and the required investments for integrated disease and pest management systems. The prospects of using modern biotechnological tools to improve cocoa, and how these approaches can reduce the negative impacts of current protection measures on the ecology and production systems are highlighted. Key recommendations have been made for all stakeholders in the cocoa industry to ensure future sustainability.

Keywords: chocolate industry, pricing, small-scale farmers, cocoa diseases, management, biotechnology

1. Introduction

Cocoa is an important crop belonging to the genus *Theobroma* in the family Malvaceae. Species of this genus are found in the wild of the Western Hemisphere rain forest from 18°N to 15°S [1]. The cultivation of cocoa, *Theobroma cacao*, L., started in equatorial Mexico between 1500 and 400 BC, and the beans were first consumed as a drink by the Mayans and Aztecs [2]. This was confirmed in earlier classical work on cocoa by van Hall [3], who emphasized that cocoa had been revered as "food for the gods," an important cultivated crop not only consumed by the native Indians as a beverage but used as a substitute for money. The revered status of cocoa was enshrined in the latinized name of "*Theobroma*" which was assigned by the botanist Linnaeus. Recent archeological records suggest that plantation-scale cultivation of cocoa occurred in the lowlands of the Mayas in the state of Chiapas and Western Belize centuries before the arrival of the Spanish [4, 5].

1.1 Spread of cocoa

Cilas and Bastide ([5], p. 2) provide a graphical timeline of the cultivation and transportation of cocoa, which began in the 1200s in Central America and then to Southern America in the sixteenth century. This was followed by cultivation in South East Asia in the first half of the 1800s and finally to the humid tropical countries of West Africa in three successive events that started in 1822. From these events, cultivation of cocoa spread through all the humid tropical lowlands and is now grown in 57 countries on three continents [6, 7]. See **Figure 1**: map of global cocoa-growing countries below [8].

1.2 Current production of cocoa

Cocoa production has rapidly increased over the past 40 years to a total of 4.9 million tons in the 2021/2022 cocoa season, with 75% coming from West Africa [9]. Côte d'Ivoire and Ghana alone account for 63% of the total global production (See **Figure 2** below).

It is evident from the graph that the sustainability of global cocoa production depends on how the West and Central African countries are able to deal with the existing and future threats, particularly regarding pests, diseases and climate change.

1.3 History of cocoa cultivation

Increased popularization of cocoa began in 1592, when the Spanish explorer, Hernan Cortés introduced cocoa drink in Spain. To reduce the bitter taste, he added sugar to the drink. It was here it became accepted as a beverage and was taken to

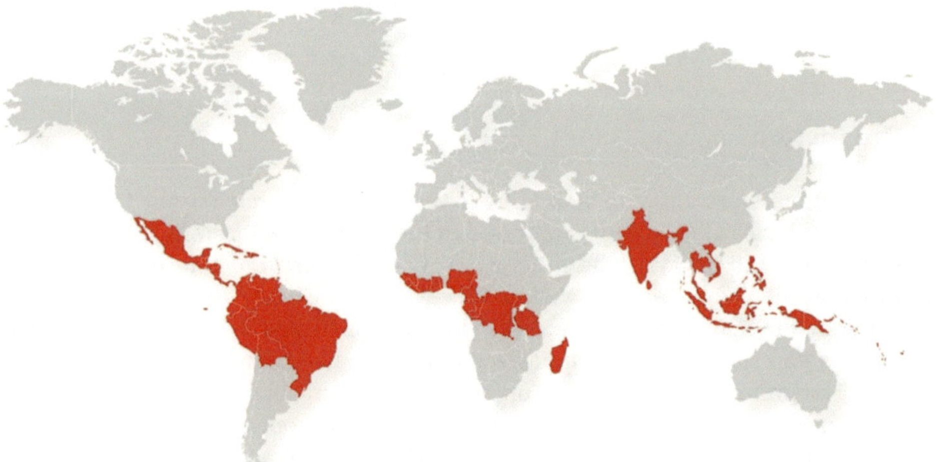

Figure 1.
Cocoa growing countries of the world (source: ICCO, 2023).

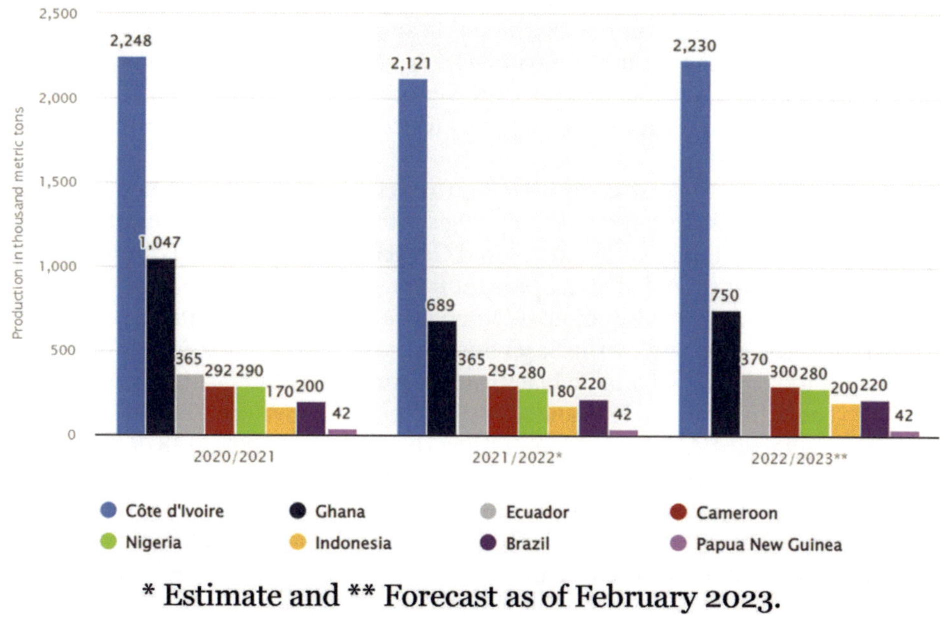

Figure 2.
Global cocoa bean production (in 1000 metric tons) by top eight countries from 2020/21 to 2022/23 (source: Statista, 2023).

other European countries, including Italy, France, Belgium and England [10]. The developed taste for chocolate beverages in Europe sparked cocoa trading. It opened the way for large-scale cultivation of cocoa by European slave merchants in plantations in the West Indies in the late seventeenth century; and Central and South America (e.g., Mexico, Venezuela, Ecuador and Surinam). Cocoa was planted in Brazil much later in the 1780s. The powerful economic gains from the production and export of cocoa beans into Europe to feed the appetites of royals, aristocrats and the growing middle class fuelled the expansion of cocoa cultivation in the fertile tropical climes.

However, the cultivation of cocoa in Africa began much later. In 1822, the Portuguese transported cocoa from Brazil and established plantations in Principe and Sao Tome in West Africa. Principe and Sao Tome became the fourth largest production region exporting over 34 million kilograms in 1901 ([3], p. 34). Later in the century, seeds were taken to Ghana, Nigeria and Côte d'Ivoire to form the basis for cocoa growing in West Africa [10].

Unfortunately, cocoa production in the plantations of Principe and Sao Tome by the Portuguese planters was largely achieved with the inhumane system of black slave labour, which was very different from the serviçal labour system laid by Royal decrees of the ruling Portuguese elites [11]. The slave labour policy on the cocoa plantations in Principe and Sao Tome followed the earlier transatlantic exportation of African slaves to the Portuguese Americas to plug in for the declining availability and use of Indians as slave labour on the sugar plantation, which prevailed in the sixteenth and seventeenth centuries [12]. The author contends that the deferential Portuguese slave labour policies amongst Indians and Africans on both sides of the Atlantic Ocean were influenced by the religious ideologies that considered labour as God's punishment for Adam's sin. Therefore, it was justified to use unfree slave labour as "the hands and feet of the noble sugar-mill masters" ([11], p. 284). When the Jesuits became mills and farm owners, they ideologically justified the enslavement of blacks but fought against the captivity of the Amerindians on the basis that they had souls just like the whites and could be used to control the black African slaves.

Contrary to the Portuguese colonizers approach in Sao Tome and Principe, where large plantations were established using captive African slave labour from Angola, cocoa cultivation in Ghana (the Gold Coast) and the rest of Africa followed a completely different model of small peasant farmer plantations ([3], p. 8). This was because earlier attempts by colonial Dutch missionaries, who planted cocoa in the coastal areas in 1815, and by the Basel missionaries at Aburi later in 1857 had both failed. It took a local Ghanaian blacksmith named Tetteh Quashie, who brought Amelonado cocoa beans from Fernando Po (Equatorial Guinea) and successfully planted a cocoa farm at Akwapim Mampong in the Eastern Region in 1879 [13]. He later sold seed pods to other local farmers, who showed interest in cultivating the crop, which spread cocoa in the region. In 1886, the colonial British Governor seeing the potential, imported cocoa pods from Sao Tome, raised seedlings at the Aburi Botanical Garden and distributed them to farmers [13]. Cocoa was cultivated in Nigeria in 1874, Cameroon in 1876, Côte d'Ivoire in 1919 [14].

1.4 Conditions for growing cocoa

Cultivated cocoa is largely divided into two subspecies Criollo and Forastero, with the latter divided into several varieties [10]. The Criollos, dominate Central America and are characterized by rounded beans, white in cross-section and a special weak flavor. On the other hand, the Forasteros have smaller and flatter beans with violet cotyledons. They have higher fat content and stronger flavor that provides the basis for plain and milk chocolate. Amelonado, a Forastero variety, was first grown in Brazil and Ecuador and later in Fernando Po in West Africa, from where it was spread to other countries in the subregion in the late nineteenth century ([10], p. 4).

Several factors influence cocoa health such as soil type and fertility, the amount of rainfall, humidity, wind and shading, crop management and pest and disease control. Wood and Lass [10] summarized the optimal growth conditions for cocoa as follows: rainfall of 1250–3000 mm per annum and preferably between 1500 and 2000 mm, a dry season of no more than 3 months; mean maximum temperature between 30 and 32°C; mean minimum of 18–21°C, and an absolute minimum of 10°C and no strong winds. Sale [15], using controlled-environment growth rooms, showed that cocoa functioned satisfactorily with high humidity (80–95% RH) at about 27°C.

Generalization about a good or suitable cocoa soil is very difficult, since soil types and conditions tend to vary significantly from one country to another. Likewise, different countries, for instance, Côte d'Ivoire and Nigeria emphasize on physical texture, based on analytical data, while in Ghana good cocoa soils are said to be deep, vary from loamy sands to friable clays, red or reddish-brown in color and should have a pH greater than 6.0 [16]. Contrarily, cocoa was successfully grown on heavy clay soil, yellow to red overlying a deposit of hydrated iron oxides in Democratic Republic of Congo [10]. Nevertheless, several cocoa soils analyzed from different countries tended to fall into the alfisols and ultisols classification [17].

2. Global economic and social impact of cocoa

Consumption of cocoa continues to grow and impacts the world economy, particularly in the emerging middle-income countries such as Brazil, China, India, Mexico and in Eastern Europe [18]. Revenue from the sales of cocoa beans significantly influences the GDP of many producing countries, particularly Côte d'Ivoire, Ghana, Cameroon, and Nigeria [8, 19]. Currently, cocoa accounts for 40% of Côte d'Ivoire, the world's largest producer's GDP. Both Ghana and Côte d'Ivoire have experienced significant deforestation of their primary forests, which is of great concern for sustainable production [20]. Likewise, there are serious issues of exploitation and use of child labour in the cocoa production processes [2].

Of the total annual cocoa production processed into cocoa mass, cocoa butter, cocoa powder, chocolate or other products, over 79% takes place outside the main producing centres in Africa [9]. Over one-third (36%) of the beans are processed in Europe, followed by Asia and Oceania (23%), then Africa (21%) and the Americas (20%). A comprehensive evaluation of global cocoa production states that the chocolate industry surpassed a retail value of USD 100 billion in 2021 and is expected to grow at a compound growth annual rate of 4.5–5.7% until 2027 [21].

These figures show the enormous contribution of the cocoa industry to the global economy. Unfortunately, trade liberalization reforms undertaken by the producing countries in the 1980s and 1990s have concentrated power in the hand of a few transnational companies at the expense of the small cocoa farmers, who are the pillars of the industry [22]. Farmers and their producing countries have suffered low farmgate and producer prices respectively, due to the unjustifiable imbalance of power at the hands of the controlling buyers and grinders, and the transnational chocolate companies and retailers, who enjoy 80–90% share of margins generated from cocoa products [21]. This immoral situation, where the hardworking farmers are just reduced to "price takers," with no bargaining power have left one in three (of the estimated 6 million cocoa farmers) in poverty, without adequate financial resources to take care of their families or invest in integrated crop management practices that could ensure their farms remain healthy and contribute to the sustainable future of the industry ([21], p. 17).

3. Impact of pest and diseases

Of the many challenges cocoa farmers face, diseases remain the most serious constraint to economic production. Available reports estimate that total global cocoa bean losses due to the major pest and diseases stands at 1 million tons, which is between 30 and 40% of the annual production [7]. For over a century, diseases have continued to pose a major threat to cocoa production due to lack of durable resistant cultivars. This is exacerbated by lack of well-funded technical infrastructure in terms of effective extension services support to farmers [7, 23]. According to Appiah [17], the ranking of cocoa diseases due to their severity, impact as limiting factor(s) to profitable production, and regional importance in the 12 leading producer countries are as follows:

3.1 *Phytophthora* pod rot (black pod disease)

Several *Phytophthora* species infect cocoa, causing leaf blight, bark canker and pod rot diseases [24]. Black pod disease is found in all the cocoa-growing regions of the world, and in particular, West Africa, where it is most severe. In 1985, worldwide losses were estimated at £1540 million [25]. Van der Vossen [6] reported that black pod disease causes an estimated 44% of the total global crop loss. More recently, Bowers *et al.* [26] stated that global black pod losses were $423 million. It is evident from these figures that black pod has become increasingly a major concern to global cocoa production.

Of the major species, *Phytophthora megakarya* (Brasier & Griffin) is indigenous to West Africa [14] and it is the most aggressive. *Phytophthora palmivora* (Butl.) Butl. is ubiquitous, *P. capsici* (Leonian) is found in South & Central America, West Indies and India and *P. citrophthora* (Smith & Smith) Brazil, Mexico and India [27]. The dynamics of *Phytophthora* infections in West Africa has changed dramatically over the years. For example, until the mid-1980s, only *P. palmivora* was known as the causal agent of the disease in Ghana. Crop losses attributed to this species were estimated at between 4.9 and 19%. However, in 1986, *P. megakarya*, was identified in the Ashanti Region of Ghana, which caused severe crop losses ranging between 60 and 100% [28]. Nationwide surveys showed that *P. megakarya* had spread rapidly across the country and threatened the livelihood of many cocoa farmers [29]. *P. megakarya* has become the dominant species and has spread west-wards to all West African cocoa-growing countries beginning from Cameroon, where it is predominant [30] through Nigeria, Togo, Ghana and Côte d'Ivoire, and southward to Gabon and Equatorial Guinea [24, 31].

3.2 Witches' broom disease

Witches' Broom disease caused by *Moniliophthora perniciosa*, is the most threatening cocoa disease in Central and South America. The disease begins when fungal spores germinate and infect meristematic tissues, developing into biotrophic hyphae that slowly occupy the intercellular spaces causing hypertrophic growth of buds, which gives the characteristic witches' broom from which the name is derived [32]. It also causes pod infection, which can lead to a high percentage of pod loss. The disease causes an estimated 29% of global crop loss [33]. Witches broom is restricted to the Western Hemisphere, including Central and South America and the Caribbean. It is currently a limiting factor to cocoa production in several Latin American countries [10]. The fungus is indigenous to the Amazon Basin. A significant spread occurred in 1984, when the disease was detected in the traditional cocoa-growing State of Bahia, Brazil, which then produced over 300,000 tonnes of cocoa per annum [10]. Currently, annual pod losses due to the disease reach between 50 and 90% in many parts of the Amazon region and production declined to 185,000 tonnes in 1997 [34].

3.3 Cocoa swollen shoot virus disease (CSSVD)

CSSVD is caused by a virus [35]. It has been and is still a major problem of all the cocoa-growing countries in West Africa [33], particularly in Ghana, where very virulent strains led to the removal of millions of Amelonado trees and were replaced with tolerant Upper Amazon hybrids [35]. The CSSVD outbreak was first reported in Ghana, then Liberia and Sierra Leone, followed by Nigeria, Côte d'Ivoire, and Togo [33]. CSSVD causes 11% of global crop loss [6]. There are many strains of the cocoa swollen shoot virus, which differ in the symptoms they produce, the vectors that transmit them and the range of their alternative hosts [10]. Virulent strains predominate and cause various types of leaf chlorosis, root necrosis, root and stem swellings and dieback in cocoa. It is quite unfortunate that after eight decades of CSSVD control and research, there are still no resistant cultivars available for farmers, the eradication and replanting policies have not been implemented properly, and new infections continue in the Western Region, which is the most concentrated cocoa growing area of Ghana [35].

3.4 Vascular-streak dieback (VSD)

VSD is caused by a basidiomycete originally named *Oncobasidium theobromae* but now *Ceratobasiduim theobromae* [36]. The pathogen causes streaking of the vascular tissue and yellowing of one or two leaves in the second or third flush from the growing tip with a characteristic pattern of green spots scattered over the yellow background. Infected leaves fall within a few days of yellowing, and the infection spreads to neighboring leaves. This leaves a distinctive situation where the youngest and oldest leaves are present, but all the middle ones are fallen. This distinguishes infection from physiological dieback due to environmental stress or insect attack [36]. VSD causes 9% of the total global loss and is important, particularly in Indonesia, Malaysia, and Papua New Guinea [6].

3.5 Moniliophthora pod rot

Moniliophthora pod rot, popularly known as frosty pod rot due to the frosty appearance of the white mycelial mat, is caused by *M. roreri* and infects only green pod tissues [37]. The pathogen grows between the parenchyma cells of the cortex, covering the pod with a white mycelial mat after the lesions have coalesced and produced conidia both within and on the surface of the host tissue [37]. It causes an estimated 5% of the total global cocoa losses and is an increasingly serious problem in Ecuador, Colombia and Central America. Frosty pod disease is said to be the most difficult to control because of the environmental resilience of its spores, ease of spread, profuse sporulation on affected pods and latent infection that can be transported great distances before conspicuous symptoms develop as well as great susceptibility of cocoa to the disease [27]. The threat posed by this disease to other continents, especially Africa, is becoming more apparent due to the vast numbers and persistence of its conidia, as well as its ability to be dispersed by wind.

It is evident from the above that diseases and pests pose a real threat to the sustainable production of cocoa. This is demonstrated by how Ghana lost the leading producer position to Côte d'Ivoire due to the CSSV epidemic that led to the destruction of a large population of cocoa trees. Similarly, disease and pests have drastically reduced cocoa production in Brazil and Malaysia [5] to the extent that Malaysia has become a net importer of cocoa.

4. Environmental and climate change impact

According to Cilas & Bastide [5] climate change is having a significant impact on land and water availability for cocoa production, changes in wind, increase in temperatures and carbon dioxide levels are contributing to higher mortality of young trees and the spread of disease vectors. Lahive *et al.* [38] predicted that atmospheric CO_2 concentration would rise to about 700 ppm by 2100. They show that such enhanced carbon dioxide level has a positive impact in stimulating photosynthesis in both cocoa seedlings and mature cocoa trees, and it appears to ameliorate some of the negative effects of water deficit through improvements in water use and quantum efficiencies.

However, the long-term effect of increased CO_2 occurred in pod biomass instead of bean dry weight per pod, which was not significantly affected. The authors suggested that an alteration in biomass allocation patterns occurs under enhanced CO_2 conditions. This could be a physiological response to adverse factors, such as water stress and temperature increases. Climate change would have an impact on cocoa's response to new pests and diseases and the spread of existing ones. These areas require further research.

According to Laderach *et al.* [39]'s climate modeling, although there would be a relatively drastic decrease in the climatic suitability of the current cocoa growing areas in Ghana and Côte d'Ivoire due to predicted increases in temperature up to 2.0°C, and decreases in monthly precipitation by 2050, there was no need for panic. Instead, we should focus on measures that would reduce the vulnerability of cocoa farmers. These include breeding more drought-resistant cocoa varieties, encouraging crop diversification, conduct research into management practices that would make farms more resilient to increasingly severe and frequent dry spells. We should adopt spatially differentiated communication and engagement strategies that would allow stakeholders evolve appropriate adaptation measures based on their geographical circumstances.

4.1 Integrated disease management strategies

In conjunction with other agronomic and soil factors, healthy cocoa production requires effective integrated pest and disease control measures [7, 40]. These should include determining the appropriate shade regimes, regular weeding, fertilization, pruning, sanitation (removal of diseased materials), timely application of environmentally friendly and target-specific pest and disease chemicals, use of locally identified biological control agents, and importantly, planting of improved disease tolerant or resistant cultivars. The effectiveness of the management and control of cocoa diseases in different countries vary in terms of techniques as well as in level of efficacy. This is due to several factors including the variation of disease-causing pathogens, availability of extension services support, phytosanitary practices and the climatic conditions involved [17].

4.2 Phytosanitary control

Cultural control strategies are environmentally friendly measures and are generally aimed at reducing humidity and the sources of inoculum for infection on the cocoa farm. A summary of cultural control practices employed in *Phytophthora* pod rot control is presented in **Table 1** [from [17]].

In general, management practices increase aeration, which help in black pod disease control. In Bahia, Brazil, shade reduction in commercial plantations lowered the incidence of *P. palmivora* black pod disease by around 40% [41]. In Ghana, a package of phytosanitary practices has been shown to adequately control black pod disease caused by *P. palmivora* [40]. These methods are less expensive and potentially affordable to small-scale farmers compared with the cost of chemical control but are time-consuming.

Activity	References [Cited from [17]]
Good aeration and exposure to sunlight	Muller [1974]
Regular weeding and judicious reduction of overhead shade	Newhall & Diaz [1967], Dakwa [1973]
Removal of mistletoes and basal chupons Ameliorative pruning of canopy	Dade [1927]
Planting at recommended spacing and draining of stagnant waters	
Regular and frequent harvesting	West [1936], Owen [1951], Hislop [1964]
Removal of diseased and mummified pods*	Thorold [1953]

**Considered unimportant by Dade [1927] and West [1936].*

Table 1.
Cultural practices used in black pod disease control.

4.3 Chemical control

The spraying of protective fungicides has been practiced for over 50 years to minimize black pod loss, but this has not always been economical [40]. Black pod disease control with chemicals primarily involves spraying protective fungicides with a pneumatic knapsack sprayer to coat healthy pods. Different copper-based protective fungicides have been tested and are used in black pod disease control. These include Bordeaux mixture (copper sulphate and calcium hydroxide, 25.43% Cu *a.i.*), Kocide 101 (77% cupric hydroxide *a.i.*) and Copper Nordox (50% cuprous oxide *a.i.*). Protective fungicides must be sprayed frequently for effectiveness due to their mode of operation and problems associated with pod growth and the cocoa environment. This requires technical training and involves high labour and input costs, which many subsistence farmers are unable to afford. There are situations where trees are so tall that fungicide sprays are not able to reach the canopy.

Copper has been shown to be redistributed in water [42]. Peirera [43] took advantage of this phenomenon in his single application technique developed against *P. palmivora* in Brazil. The method involved spraying higher doses of fungicide, which is later washed down the trunk to effect control. On the same basis, Sreenivasan *et al.* [44], tied materials impregnated with copper fungicide about two metres from the base of cocoa trees which was redistributed slowly down the tree by rain water.

These methods work against *P. palmivora* black pod but are not effective against *P. megakarya*, due to the vast differences in the sporulation abilities and their main sources of primary inoculum (tree canopy for *P. palmivora* and soil for *P. megakarya*). Opoku [45] compared the production of sporangia and zoospores by *P. megakarya* and *P. palmivora* on different media and established that *P. megakarya* produced 4–6 times more sporangia and zoospores than *P. palmivora* under all conditions.

A semi-systemic fungicide, Ridomil 72 Plus (60% cuprous oxide, 12% metalaxyl), and a single injection of Foli-R-Fos 400 (potassium phosphonate, 40% *a.i*), which is a systemic fungicide and a foliar fertilizer has shown to be more effective than contact copper fungicides in the control of both *P. megakarya* and *P. palmivora* [46]. Studies in Ghana have also shown that extended intervals of four-weekly applications using Nordox 75 and Ridomil 72 Plus effectively and economically control black pod disease caused by *P. megakarya* [47]. The four-weekly spraying significantly reduces the number of applications per cocoa season (May to October) from the current recommendation 8–9 to 5–6 and communicated to farmers could reduce production cost and increase adoption. Phosphonic acid is not subject to any patent, has lower toxicity compared to contact fungicides and the single injection provides lower economic cost, lower operator risk and no environmental contamination of the cocoa ecosystem.

In Central and South America, copper-based fungicides and azoles are used to control witches' broom and frosty pod diseases, albeit not very effective due to many constraints enumerated above [48]. Generally, the profitability of fungicide application depends on the level of farm management, the nature of land tenure and labour arrangements for farm operations [40]. However, due to increasing concerns about antifungal resistance and negative impacts on human health and the environment, alternative strategies are desperately needed [49].

4.4 Biological control

Peirera [50], in a review of prospects for effective control of cocoa diseases, mentioned that while the textbook advantages of biological control management strategy are not in doubt, in cocoa, the promise of actual field use has not been realized, and more research is required. Antagonism *in vitro* of *P. palmivora* by biocontrol agents has been demonstrated. Galindo [51], reported that *Pseudomonas fluorescens* isolated from the surface of a healthy cocoa pod was antagonistic to *P. palmivora in vitro* and was more effective than copper oxide and chlorothalonil in the field. More recent biocontrol efforts have shown mixed results. In Peru, Kraus and Soberanis [52] showed that *Trichoderma* spp. reduced moniliasis, witches' broom and black pod. However, in Costa Rica, field application isolates of the hyper parasitic fungi *Clonostachys byssicola* and *Trichoderma asperellum* made no significant improvement to healthy yields [53]. Ferraz *et al.* [48] touted the potential of yeast species as they are safe for humans, easy to manipulate, shown to enhance plant wellbeing and being environmentally friendly biocontrol agents against witches' broom disease in Brazil.

In Africa, Deberdt *et al.* [54] found that *Trichoderma asperellum* biocontrol agent (strain PR11) was promising in Cameroon but not as effective as the fungicide treatment under high disease pressure. Therefore, integrating biocontrol agents into an IPM strategy was recommended. In Nigeria, farmers have a favorable disposition towards the use of bioagents due to high cost and safety concerns of synthetic fungicides [55]. Similarly, biological control efforts have been made in Côte d'Ivoire, Kebe *et al.* [56] reported that isolates of *Trichoderma* sp. showed fungicidal effects; and two bacteria isolates of the *Bacillus* genus significantly reduced cocoa leaf susceptibility to *P. palmivora*. In Ghana, Akrofi *et al.* [57], in a study of cocoa microbiota obtained 17 isolates, mainly *Pseudomonas* species from three notable cocoa varieties in Ghana. These demonstrated significant inhibition of mycelia growth of *P. palmivora* on plates and prevented disease establishment on pods.

The above results show potential and receptivity to biocontrol as a better alternative to the prevailing copper-based fungicides, which have non-target, food chain contamination and environmental effects [50]. Therefore, serious research investments and greater efforts in the producing countries are needed to find effective biocontrol agents against the endemic pathogens causing cocoa diseases in the different geographic sub-regions. The search for biocontrol agents should focus on the areas identified as the centres of origin or diversity of each cocoa disease, since the potential to find co-evolved natural control agents are high in these areas. A fully integrated pest management approach that utilizes all the available methods including endophytes and mycoparasites, development of tissue culture and tolerant cultivars, should be pursued to minimize the application of fungicides, which the chocolate industry is tightening control of their use on health grounds, as residues contaminate the food chain and their unintended impact on the cocoa ecosystem.

4.5 Development of resistant varieties

Developing genetic resistance against the five major diseases of cocoa have a long history beginning from the Pound collections in 1938, aimed at selecting and accumulating desirable characteristics including high-yielding and disease resistance or tolerance varieties [58]. It is undeniable that genetically resistant varieties are the most cost-effective and reliable method of disease control [59]. Over the years, different breeding and screening techniques have been developed with significant success and challenges (for details see [58]).

In 1978, Lawrence [60] stated that no major genes for resistance to *P. palmivora* had been identified in *T. cacao*. Ten years later, Phillip-Mora and Galindo [61] reported 19 resistant cocoa cultivars to *P. palmivora* in Costa Rica. In Ghana, field evaluation of individual cocoa trees for resistance to *P. megakarya* began in 1990 in two endemic areas: Bechem and Akomadan in the Ashanti Region. From 25 trees that were selected as "resistant" parents, nine showed promise against *P. megakarya* after the challenge inoculation of attached pods. The high level of susceptibility obtained was attributed to the narrow genetic base of the parents [62]. However, in 1997, Van der Vossen [6] noted that no genotype had been found with complete resistance to black pod diseases caused by either *P. palmivora* or *P. megakarya*.

There are many challenges in cocoa breeding work. These include narrow genetic base of available germplasm, annual nature of cocoa production, differences in the diseases caused by the same species, the presence of multiple diseases in the same sub-region and the geographic separation in the areas of influence of each major disease. For example, each of the five *Phytophthora* species involved in black pod disease differs in structure, geographic distribution, ecology and pathogenicity. These challenges necessitate that prospective new cultivars have to be tested for resistance to each of the disease pathogens present. This had not been possible until recently, due to lack of international collaborative projects between producing countries and the danger of introducing new diseases due to poor and inadequate quarantine facilities.

The CFC/ICCO/IPGRI[1] project [63], which involved 10 major cocoa-growing countries and international centres for cocoa germplasm conservation and improvement, addressed these barriers. The objectives were to select better cocoa varieties, reinforce population breeding activities, characterize, evaluate, and enhance cocoa germplasms with emphasis on disease and pest resistance. Twenty-five locally selected clones were tested against local isolates of pathogen and "ring tests" involving isolates from participating countries against the selected germplasm in a non-cocoa growing country. According to Eskes [64], the follow-up project evaluated 1500 trees selected by farmers for yield or low disease or pest incidence that are high-yielding and pest- and disease-resistant candidates, which have been released to cocoa farmers. This demonstrates the power of global collaborative research and the practical results from large investments in research.

The challenges of standardized protocols in accessing levels of resistance expressed in different tissues used in screening work were addressed by the development of the leaf disc inoculation technique [65]. Subsequently, Iwaro *et al.* [66]

[1] CFC/ICCO/IPGRI – Common Fund for Commodities/International Cocoa Organization/International Plant Genetic Resources Institute

demonstrated a strong correlation between detached and attached leaf lesions. They also assessed the resistance to *P. palmivora* in leaves and pods of different genotypes at the penetration and post-penetration stages of infection. A significant correlation between the resistance of leaves and pods at the post-penetration stage was established, showing that internal resistance is common between leaves and pods and that leaf resistance at this stage could be used to predict pod resistance. A high positive correlation between attached leaves and pods with their detached counterparts was confirmed. These screening techniques are now accepted standard tools for early screening. Different mechanisms may be involved in penetration and post-penetration resistance [17, 66]. Depth of inoculation and stages of pod maturity influenced the level of resistance; hence there is a need to standardize these factors in the screening of cocoa germplasm for resistance to *Phytophthora*.

Expression of defense gene response to *P. palmivora* in different genotypes of cocoa has been found to occur in blocks. These are constitutively expressed at different levels and are potential sources for the many quantitative trait loci (QTLs) contributing to resistance in cocoa [67]. Selecting host resistance is needed for all the major diseases of cocoa. However, the lack of resistance in the international cocoa germplasm collection (ICGC) is a major challenge [68]. It is also interesting to note that some clones in the ICGC have been designated as "resistant" and also susceptible to isolates of the same pathogen. This illustrates the variability in pathogens and the lack of durability, as a clone could be resistant to one isolate and susceptible to another. Both CSSVD and VSD are systemic diseases, and a better understanding of the mechanism of resistance is needed. According to Dzahini-Obiatey *et al.* [35], the ultimate strategy to overcome viral diseases will be to produce genetically engineered cocoa plants by introducing resistant genes using non-conventional biotechnological techniques.

4.6 Modern biotechnological tools: Hope or hoax?

Recent technological advances in biotechnology offer hope for achieving long-term durable resistance in cultivated crops. The increasing legislative constraints on the use of agrochemicals and climate change challenges [49] make biotechnological solutions more imperative for agricultural crops-dependent countries. This is particularly important in the tropics, where the challenges to crop production are most severe. Governments in these countries should consider investing heavily in modern biotechnological tools and the associated capacity building to accelerate improvements in yields and, in particular, resilience to biotic (pest and diseases) and abiotic (climate change) stresses.

Considerable achievements have been made in cocoa breeding and selection, including sequencing of two cocoa genomes [69], which has allowed the identification of genes and proteins that code for specific traits [70], gene discovery and marker-assisted breeding using single-nucleotide polymorphism identifications [71], QTLs mapping of resistance [71] and genome-wide characterization [72]. These developments show great prospects towards targeted gene transfer and guided selective breeding for durable resistance in cocoa against these challenging diseases.

Currently, there are many new powerful plant biotechnological techniques, for example, genome editing, which involves precise modification of specific DNA sequences using three protein-dependent DNA cleavage systems, namely the zinc-finger nucleases, transcription activator-like effector nucleases, and RNA-dependent DNA cleavage systems (i.e., CRISP-associated proteins) [73]; RNA interference, and cisgenesis - the single gene transfer from sexually compatible crosses [74]. Amongst these modern gene editing technologies, CRISPR/Cas9 is reported to be faster, cheaper, precise, and highly efficient in editing genomes, even at the multiplex level [75]. It has been successfully used in cocoa leaves and cotyledon cells to delete the *TcNPR3* gene, which suppresses the plant defense response using *Agrobacterium*-mediated transient transformation, elevated expression of downstream defense genes and increased resistance to infection caused by *Phytophthora tropicalis* [76].

Now we have valuable genome datasets and functional tools for rapid and target-specific genetic changes that could confer vigor, resistance to diseases, pests and resilience to abiotic stresses in cocoa. These new tools for characterization of cocoa genes and genetic manipulation of disease resistance in other important tropical crops overcome problems associated with traditional breeding techniques. However, whether these biotechnological improvements would be seen as genetically modified organisms, which are not widely accepted in the northern hemisphere, remains unclear [77]. Also, these cutting-edge approaches require specialist training and expensive equipment and supplies, which many producer countries are not equipped with.

5. Conclusions

The cocoa industry, particularly chocolate manufacturers, has enjoyed a steady supply of cocoa beans that have allowed uninterrupted manufacturing of consumable products and the holding of good buffer stocks. Conversely, cocoa producers continue to struggle due to low prices, diseases, pests and a rapidly changing climate. Cocoa has been described as a crop "produced in the south and consumed in the north" [9]. The story of cocoa production requires rebalancing the colonial power dynamics in the value chain to bring equitable benefit to all involved.

Currently, cocoa production faces significant but not existential challenges. These include threats posed by pests and diseases, low cocoa prices paid to farmers, as well as climate change and its potential adverse impacts on food production in cocoa growing areas. The economic benefits, enjoyment and health value of cocoa products to the masses, producing countries, the large transnational businesses and the small-scale farmer (who is at the heart of everything) should serve as strong incentives for all stakeholders to come together to address the critical issues ([21], p. 6). It behooves on the chocolate industry, the largest beneficiaries to keep the goose that is laying the golden eggs healthy and productive. There is a need for investment in critical research and development, adoption of fair and ethical trading policies. An introspective look and re-examination of the power imbalances in the existing cocoa value chain are urgently needed for a sustainable cocoa future.

5.1 Recommendations

The quest for sustainable cocoa production that adopts environmentally friendly farm management practices, avoids child labour and embraces the voluntary standards would happen if farmers are given their deserving living income from their work [21, 77]. To address the perennial threat of pest and diseases, increased efforts and heavy investments are required to develop cocoa varieties that are truly resistant. We propose several interrelated political and technical recommendations to key stakeholders. These are:

5.2 Governments of coca producing countries

The governments of producer countries should:

- Strengthen the Cocoa Producers Alliance to become a strong and united organization that could control production levels (like the Organization of Petroleum Exporting Countries), negotiate just a share of cocoa income in terms of producer prices, and adopts uniform trading policies.

- Engage in serious dialog with the powerful Western cocoa trading and chocolate manufacturing companies to address the existing trading imbalances.

- Develop good infrastructure including transportation systems in the rural cocoa farming areas for easy access to market.

- Provide fair pricing and government-backed financing with low-interest rate and flexible repayment terms.

1. **Scientists and Cocoa Boards**

- Producing countries and scientists backed by policymakers must develop integrated pest management strategies in partnership with farmers that allow them to utilize environmentally friendly phytosanitary practices, improved planting materials and biological control measures.

- The IPM strategy should allow limited and targeted use of approved fungicides at critical period of infection.

- Farmers should be trained in scientific agriculture and farming as a business including cocoa ecotourism.

2. **Chocolate Industry and Scientists**

- The chocolate industry should increase research funding and provide local scientists resources to conduct a non-conventional biotechnological cocoa improvement programme to hasten cocoa breeding and selection work.

- Research institutions should put greater emphasis on developing effective biological control of cocoa diseases and understanding the mechanisms responsible for the variability experienced in field trial.

- Regional collaborative biological control research groups and centres should be established with funding from the chocolate industry and Cocoa Producers Alliance.

3. **Farmers**

- It must be recognized that farmers hold the key to long-term, sustainable cocoa production; therefore, improving their conditions should be the primary concern of key stakeholders in the cocoa value chain. In this regard:

- Farmers should be encouraged to form strong local cooperatives.
- Farmers unions and associations must be empowered to collectively bargain for humane farmgate cocoa prices.
- Train farmers on environmentally friendly practices [7].

Acknowledgements

The author is very grateful to Miss Iris Xiang, Information Specialist at Avenues Shenzhen, for her assistance with finding key research manuscripts used in this study. The constant encouragement and support of Mrs. Juliana Asante Appiah (nee Sackey) is also gratefully acknowledged.

Conflict of interest

The author declares no conflict of interests.

Author details

Alex Asante Appiah[1,2]

1 STEAM Expert, Avenues Shenzhen, China

2 Founder and CEO, Centre for African Leadership and Excellence, United Kingdom

*Address all correspondence to: alex.appiah@calecentre.org

References

[1] Toxopeus H. Botany, types and populations. In: Wood GAR, Lass RA, editors. Cocoa, Tropical Agricultural Series. 4th ed. London: Longman; 1985. pp. 11-37

[2] Ould D, Jordan C, Reynolds R, Loftin L. The cocoa industry in West Africa: A history of exploitation. Anti-Slavery International. 2004. p. 71. Available from: https://www.antislavery.org/wp-content/uploads/2017/01/1_cocoa_report_2004.pdf

[3] Van Hall CJ. Cocoa. London: Macmillan and Company Limited; 1914

[4] Muhs DR, Kautz RR, MacKinnon JJ. Soils and the location of cacao orchards at a Maya site in western Belize. Journal of Archaeological Science. 1985;**12**(2):121-137

[5] Cilas C, Bastide P. Challenges to cocoa production in the face of climate change and the spread of pests and diseases. Agronomy. 2020;**10**(9):1232

[6] Van der Vossen HAM. Strategies of variety improvement in cocoa with emphasis on durable disease resistance. International Group for Genetic Improvement of Cocoa. 1997:9-18

[7] Hebbar PK. Cacao diseases: A global perspective from an industry point of view. Phytopathology. 2007;**97**(12):1658-1663

[8] ICCO. Growing Cocoa. 2023. Available from: https://www.icco.org/growing-cocoa/

[9] Swiss Platform for Sustainable Cocoa. Produced in the South, Consumed in the North. 2023. Available from: https://www.kakaoplattform.ch/about-cocoa/cocoa-facts-and-figures. [Accessed on 13th May 2023]

[10] Wood GAR, Lass RA. Cocoa. In: Tropical Agricultural Series. 4th ed. London: Longman; 1985. p. 620

[11] Burtt J, Horton WC. Report on the Conditions of Coloured Labour Employed on the Cocoa Plantations of S. Thomé and Principe and the Methods of Procuring it in Angola. Temporary American Committee of the Anti-Slavery and Aborigines Protection Society. 1907

[12] Botelho TR. Labour ideologies and labour relations in colonial Portuguese America, 1500-1700. International Review of Social History. 2011;**56**(S19):275-296

[13] COCOBOD. The Ghana cocoa story. Available from: https://cocobod.gh/cocoa-story 2023. [Accessed on 13th May 2023]

[14] Akrofi AY. Phytophthora megakarya: A review on its status as a pathogen on cacao in West Africa. African Crop Science Journal. 2015;**23**(1):67-87

[15] Sale PJ. Growth and flowering of cacao under controlled atmospheric relative humidities. Journal of Horticultural Science. 1970;**45**(2):119-132

[16] Charter CF. Cacao Soils: Good and Bad. Tafo, Ghana: (Mimeo) West African Cocoa Research Institute; 1947

[17] Appiah AA. Variability of Phytophthora species causing black pod disease of cocoa (Theobroma cacao L.) and implications for assessment of host resistance [Doctoral dissertation, Imperial College London (Univ. of London)]

[18] Zhang D, Motilal L. Origin, dispersal, and current global distribution of cacao genetic diversity. In: Bailey BA, Meinhardt LW, editors. Cacao Diseases: A History of Old Enemies and New Encounters. Switzerland: Springer International Publishing; 2016. pp. 1-31

[19] Tchokote J, Martin P, Nguezet DO, Onyebuchi OK. An economic appraisal of cocoa production in Cameroon: The case study of Lekie division. Organization. 2015;**6**(9):168-181

[20] UN SDG. Sustainable cocoa farming in Côte d'Ivoire: UN deputy chief notes significant progress and calls for greater international support. 2022. Available from: https://unsdg.un.org/latest/stories/sustainable-cocoa-farming-Côte-divoire-un-deputy-chief-notes-significant-progress

[21] Bermudez S, Voora V, Larrea C, Luna E. Cocoa prices and sustainability. 2022:1-42 Available from: https://www.iisd.org/system/files/2022-11/2022-global-market-report-cocoa.pdf

[22] Gayi SK, Tsowou K. Cocoa Industry: Integrating Small Farmers into the Global Value Chain. New York and Geneva: United Nations; 2016. Available from: https://unctad.org/system/files/official-document/suc2015d4_en.pdf

[23] Baah F, Anchirinah V. A review of cocoa research Institute of Ghana extension activities and the management of cocoa pests and diseases in Ghana. American Journal of Social and Management Sciences. 2011;**2**(1):196-201

[24] Appiah AA, Opoku IY, Akrofi AY. Natural occurrence and distribution of stem cankers caused by Phytophthora megakarya and Phytophthora palmivora on cocoa. European Journal of Plant Pathology. 2004;**110**:983-990

[25] Evans HC, Prior C. Cocoa pod diseases: Causal agents and control. Outlook on Agriculture. 1987;**16**(1):35-41

[26] Bowers JH, Bailey BA, Hebbar PK, Sanogo S, Lumsden RD. The impact of plant diseases on world chocolate production. Plant Health Progress. 2001;**2**(1):12

[27] Ploetz R. The impact of diseases on cacao production: A global overview. In: Bailey BA, Meinhardt LW, editors. Cacao Diseases: A History of Old Enemies and New Encounters. Switzerland: Springer International Publishing; 2016. pp. 33-59

[28] Dakwa JT. A serious outbreak of black pod disease in a marginal area of Ghana. In: Proceedings of the 10th International Cocoa Research Conference. Lagos, Nigeria: Cocoa Producers Alliance; 1987. pp. 447-451

[29] Opoku IY, Akrofi AY, Appiah AA. The spread of Phytophthora megakarya on cocoa in Ghana. In: Proceedings of the 1st International Cocoa Pests and Diseases Seminar, 6-10 November. Tafo, Ghana: Cocoa Research Institute; 1997. pp. 98-107

[30] Nyassé S, Grivet L, Risterucci AM, Blaha G, Berry D, Lanaud C, et al. Diversity of Phytophthora megakarya in central and West Africa revealed by isozyme and RAPD markers. Mycological Research. 1999;**103**(10):1225-1234

[31] Coulibaly K, Kebe IB, Koffi NK, Mpika J, Kone D. Characterization of isolates of Phytophthora spp from cocoa orchards in Côte d'Ivoire. Journal of Applied Biosciences. 2013;**70**:5567-5579

[32] Mondego JM, Carazzolle MF, Costa GG, Formighieri EF, Parizzi LP, Rincones J, et al. A genome survey of Moniliophthora perniciosa gives new

insights into Witches' broom disease of cacao. BMC Genomics. 2008;**9**:1-25

[33] Marelli JP, Guest DI, Bailey BA, Evans HC, Brown JK, Junaid M, et al. Chocolate under threat from old and new cacao diseases. Phytopathology. 2019;**109**(8):1331-1343

[34] ICCO. ICCO Quarterly Bulletin of Cocoa Statistics, London, Vol. XXIV, No. 3. Abidjan, Côte d'Ivoire: International Cocoa Organization; 1998. p. 69

[35] Dzahini-Obiatey H, Domfeh O, Amoah FM. Over seventy years of a viral disease of cocoa in Ghana: From researchers' perspective. African Journal of Agricultural Research. 2010;**5**(7):476-485

[36] Prior C. Vascular-streak die back. Cocoa growers. Bulletin. 1980;**29**: 21-30

[37] Evans HC, Stalpers JA, Samson RA, Benny GL. On the taxonomy of Monilia roreri, an important pathogen of Theobroma cacao in South America. Can. Journal of Botany. 1978;**56**(20):2528-2532

[38] Lahive F, Handley LR, Hadley P, Daymond AJ. The impacts of climate change variables on vegetative and reproductive development of six genotypes of cacao. In: International Symposium on Cocoa Research (ISCR). Lima, Peru, Abidjan, Côte d'Ivoire: International Cocoa Organization; 2017. pp. 13-17

[39] Läderach P, Martinez-Valle A, Schroth G, Castro N. Predicting the future climatic suitability for cocoa farming of the world's leading producer countries. Ghana and Côte d'Ivoire. Climatic change. 2013;**119**(3-4):841-854

[40] Akrofi AY, Appiah AA, Opoku IY. Management of Phytophthora pod rot disease on cocoa farms in Ghana. Crop protection. 2003;**22**(3):469-477

[41] Anon. Environmental factors associated with black pod disease of cacao. Revisita Theobroma, CEPEC, Itabuna-Brasil. 1972. Ano 2: 34 (abstract)

[42] Hislop EC, Cox TW. Local redistribution of fungicides on leaves by water. Annals of Applied Biology. 1970;**66**(1):89-101

[43] Pereira JL. Chemical control of Phytophthora pod rot of cocoa in Brazil. Cocoa Growers' Bulletin. 1985;**36**:23-38

[44] Sreenivasan TN, Pettitt TR, Rudgard SA. An alternative method of applying copper to control cocoa pod diseases. In Brighton Crop Protection Conference, Pests and Diseases-1990. Vol. 2. Cambridge, UK: British Crop Protection Council; 1990. pp. 583-588

[45] Opoku IY. Survival of 'Phytophthora palmivora' and 'Phytophthora megakarya' in Soil and Cocoa Roots [Doctoral dissertation], Univ. of London. 1994

[46] Opoku IY, Akrofi AY, Appiah AA, Luterbacher MC. Trunk injection of potassium phosphonate for the control of black pod disease of cocoa. Tropical Science. 1998;**38**(3):179-185

[47] Akrofi AY, Opoku IY, Appiah AA. Extended interval of application of fungicides for the control of black pod disease in Ghana. In: Proceedings of the 2nd International Cocoa Pests and Diseases Seminar, January 19-24 Yamoussoukro, Côte d'Ivoire. Lagos, Nigeria: Cocoa Producers Alliance; 1998

[48] Ferraz P, Cássio F, Lucas C. Potential of yeasts as biocontrol agents of the

phytopathogen causing cacao witches' broom disease: Is microbial warfare a solution? Frontiers in Microbiology. 2019;**10**:1766

[49] Fisher MC, Hawkins NJ, Sanglard D, Gurr SJ. Worldwide emergence of resistance to antifungal drugs challenges human health and food security. Science. 2018;**360**(6390):739-742. DOI: 10.1126/science.aap7999

[50] Pereira JL. Prospects for effective control of cocoa diseases. In: Proc. 12th International Cocoa Research Conference. Salvador, Bahia, Brasil; 1996. pp. 81-86

[51] Gallindo JJ, Keane PJ, Putter CA. Prospects of biological control of black pod of cocoa. Cocoa Pest and Disease Management in South East Asia and Australasia. 1992;**112**:31-38

[52] Krauss U, Soberanis W. Effect of fertilization and biocontrol application frequency on cocoa pod diseases. Biological Control. 2002;**24**(1):82-89

[53] Bateman RP, Hidalgo E, Garcia J, Arroyo C, Ten Hoopen GM, Adonijah V, et al. Application of chemical and biological agents for the management of frosty pod rot (Moniliophthora roreri) in Costa Rican cocoa (Theobroma cacao). Annals of Applied Biology. 2005;**147**(2):129-138

[54] Deberdt P, Mfegue CV, Tondje PR, Bon MC, Ducamp M, Hurard C, et al. Impact of environmental factors, chemical fungicide and biological control on cacao pod production dynamics and black pod disease (Phytophthora megakarya) in Cameroon. Biological Control. 2008;**44**(2):149-159

[55] Adejumo TO. Crop protection strategies for major diseases of cocoa, coffee and cashew in Nigeria. African Journal of Biotechnology. 2005;**4**(2):143-150

[56] Kebe I, Mpika J, N'guessa KF, Hebbar PK, Samuels GS, Ake S. Isolement et identification de microorganismes indigènes de cacaoyères en côte d'ivoire et mise en évidence de leurs effets antagonistes vis-àvis de Phytophthora palmivora, agent de la pourriture brune des cabosses. Sciences & Nature. 2009;**6**(1):71-82

[57] Akrofi AY, Terlabie JL, Amoako-Attah I, Asare EK. Isolation and characterization of bacteria from different cacao progenies and their antagonistic activity against the black pod disease pathogen, Phytophthora palmivora. Journal of Plant Diseases and Protection. 2017;**124**:143-152. DOI: 10.1007/s41348-017-0082-z

[58] Gutiérrez OA, Campbell AS, Phillips-Mora W. Breeding for disease resistance in cacao. In: Bailey BA, Meinhardt LW, editors. Cacao Diseases, A History of Old Enemies and New Encounters. Switzerland: Springer International Publishing; 2016. pp. 567-609

[59] Iwaro AD, Sreenivasan TN, Umaharan P. Cacao resistance to Phytophthora: Effect of pathogen species inoculation depth and pod maturity. European Journal of Plant Pathology. 1998;**104**:11-15

[60] Lawrence JS. Evaluation of methods for assessing resistance of cacao Theobroma cacao L. cultivars and hybrids to Phytophthora palmivora (Butler) Butler [Podridao parda; Variedade resistente; Cacau]. Boletim Tecnico-Centro de Pesquisas do Cacau (Brazil). no. 62. 1978

[61] Phillips Mora W, Galindo JJ. Método de inoculación y evaluación de la

resistencia a Phytophthora palmivora en frutos de cacao (Theobroma cacao). Turrialba. 1989;**39**(4):488-496

[62] Abdul-Karimu A, Opoku IY, Akrofi AY, Adu-Ampomah Y. The search for Phytophthora megakarya resistant cocoa genotypes in Ghana. II. Preliminary evaluation of some selected parents. In: Proceedings of the 2nd International Cocoa Pests and Diseases Seminar. Yamoussoukro, Côte d'Ivoire, Lagos, Nigeria: Cocoa Producers Alliance; January 1998. pp. 19, 1998-24

[63] Eskes AB, Engels JMM, Lass RA. Working procedures for cocoa germplasm evaluation and selection. In: Proceedings of the CFC/ICCO/IPGRI Project Workshop, 1-6 February 1998, Montpellier, France. Rome, Italy: International Plant Genetic Resources Institute; 1999

[64] Eskes AB. Collaborative and participatory approaches to cocoa variety improvement. Final Report of the CFC/ICCO/Bioversity project on "Cocoa Productivity and Quality Improvement: A Participatory Approach" (2004-2010). CFC, Amsterdam, The Netherlands/ICCO, London, UK. Rome, Italy: Bioversity International; 2011

[65] Nyassé S, Cilas C, Herail C, Blaha G. Leaf inoculation as an early screening test for cocoa (Theobroma cacao L.) resistance to Phytophthora black pod disease. Crop Protection. 1995;**14**(8):657-663

[66] Iwaro AD, Sreenivasan TN, Umaharan P. Foliar resistance to Phytophthora palmivora as an indicator of pod resistance in Theobroma cacao. Plant Disease. 1997;**81**(6):619-624

[67] Baruah IK, Ali SS, Shao J, Lary D, Bailey BA. Changes in gene expression in leaves of cacao genotypes resistant and susceptible to Phytophthora palmivora infection. Frontiers in Plant Science. 2022;**12**:3281

[68] Meinhardt LW, Bailey BA. Moniliophthora Roreri genome and transcriptome. In: Cacao Diseases: A History of Old Enemies and New Encounters. Switzerland: Springer International Publishing; 2016. pp. 97-135

[69] da Silva RA, Souza G, Lemos LS, Lopes UV, Patrocínio NG, Alves RM, et al. Genome size, cytogenetic data and transferability of EST-SSRs markers in wild and cultivated species of the genus Theobroma L. (Byttnerioideae, Malvaceae). PLoS One. 2017;**12**(2):e0170799

[70] Argout X, Salse J, Aury JM, Guiltinan MJ, Droc G, Gouzy J, et al. The genome of Theobroma cacao. Nature Genetics. 2011;**43**(2):101-108

[71] Mucherino Munoz JJ, de Melo CA, Santana Silva RJ, Luz ED, Corrêa RX. Structural and functional genomics of the resistance of cacao to Phytophthora palmivora. Pathogens. 2021;**10**(8):961

[72] Bekele FL, Bidaisee GG, Allegre M, Argout X, Fouet O, Boccara M, et al. Genome-wide association studies and genomic selection assays made in a large sample of cacao (Theobroma cacao L.) germplasm reveal significant marker-trait associations and good predictive value for improving yield potential. PLoS One. 2022;**17**(10):e0260907

[73] Zhang K, Raboanatahiry N, Zhu B, Li M. Progress in genome editing technology and its application in plants. Frontiers in Plant Science. 2017;**8**:177

[74] Haque E, Taniguchi H, Hassan MM, Bhowmik P, Karim MR, Śmiech M, et al. Application of CRISPR/Cas9 genome

editing technology for the improvement of crops cultivated in tropical climates: Recent progress, prospects, and challenges. Frontiers in Plant Science. 2018;**9**:617. DOI: 10.3389/fpls.2018.00617

[75] Wickramasuriya AM, Dunwell JM. Cocoa biotechnology: Current status and future prospects. Plant Biotechnology Journal. 2018;**16**(1):4-17

[76] Fister AS, Landherr L, Maximova SN, Guiltinan MJ. Transient expression of CRISPR/Cas9 machinery targeting TcNPR3 enhances defense response in Theobroma cacao. Frontiers in Plant Science. 2018;**9**:268. DOI: 10.3389/fpls.2018.00268

[77] Fairtrade Foundation. The fight for living incomes 2023. Available from: https://www.fairtrade.org.uk/get-involved/current-campaigns/the-fight-for-living-incomes/ [Accessed: 2023-05-12]

Chapter 8

The Integrated Pest Management Implementation of the Cocoa Pod Borer in Indonesia

Araz Meilin, Nurmili Yuliani, Nurhayati, Hermawati Cahyaningrum, Rein E. Senewe, Saidah, Herlina N. Salamba, Delima Napitupulu, Ismon Lenin, Risma Fira Suneth, Nur Imdah Minsyah, Handoko, Suparwoto, Endrizal, Busyra B. Saidi, Jumakir, Waluyo, Yardha, Muh. Asaad, Syafri Edi, Yustisia, Sigid Handoko, Nila Wardani and Julistia Bobihoe

Abstract

Indonesia is among the largest cocoa producers in the world and it makes an important contribution to the nation's economy. In Indonesia, the cocoa pod borer (CPB) outbreaks have caused a decline in cocoa yield and quality, impacting the livelihoods of cocoa farmers. Application of Integrated Pest Management (IPM) is promoted by the Indonesian government, in collaboration with various organizations. IPM is an approach that focuses on using a combination of biological, cultural, and chemical control methods to manage pests and reduce their impact on crops. The adoption of IPM practices in cocoa farming has shown promising results in Indonesia. Farmers have reported improved yields and quality of cocoa beans, reduced pesticide use, and improved environmental sustainability. In addition, the application of IPM has equipped farmers with knowledge and skills that can help them overcome other challenges in cocoa farming, such as a healthy environment.

Keywords: cocoa pod borer, *Conopomorpha cramerella*, Integrated Pest Management (IPM), cocoa, Indonesia

1. Introduction

Cocoa is one of the plantation commodities that support economic activities and earners of foreign exchange in Indonesia, in addition to oil and gas. In 2021, cocoa plantations in Indonesia cultivated by smallholder plantations are estimated at 1.45 million hectares (99.39%), while large private plantations cultivate 8.21

thousand hectares (0.56%) and large state plantations only cultivate 0.67 thousand hectares (0.05%). The area of cocoa plantations in Indonesia before 2021 over the last four years tends to show a decline, decreasing by around 2.55–3.33% per year. This decrease in planting area also has an impact on decreasing cocoa production [1]. Furthermore, the decline in cocoa production in Indonesia in recent years has been attributed to factors such as minimal garden maintenance, lack of technological application, and pest damage, especially the cocoa pod borer (CPB) caused by *Conopomorpha cramerella* Snellen [2].

CPB is a pest of cocoa in Indonesia for more than three decades since it was discovered in 1980 [3]. CPB infestation has been a persistent problem in Indonesian cocoa for over three decades, causing significant yield losses and economic impact for cocoa farmers. CPB infestation in Indonesian cocoa was confirmed in 1997 by Matlick [4], and the industry estimates that infestation in Sulawesi adversely affects up to 80% of cocoa farms [5]. If infestation occurs when the cocoa pod is ripe, near harvest, most of the beans in the pod remain unaffected. However, if infestation occurs when the pod is immature, its entire contents can be lost. Unfortunately, it is often difficult to detect the presence or severity of infestation. Poor harvests are experienced in Southeast Sulawesi where cocoa plantations tend to produce low-quality cocoa beans and their productivity is less than 650 kg ha^{-1} due to an inadequate cultivation system and attack by the cocoa pod borer (CPB) [6, 7]. In the Lima Puluh Kota regency of West Sumatera Province, the CPB was identified as the primary insect pest infesting cocoa pods. The percentage of cocoa plants affected by CPB was recorded at 21.18%, while the percentage of cocoa pods attacked by CPB was found to be 10.82%. The intensity of the attack by CPB was measured at 8.52% [8]. The attack of the cocoa pod borer (CPB) is a significant threat to annual cocoa production, causing yield losses ranging from 18.25% to 73.04%, as reported by [9–11]. Control measures such as good technical cultural practices during cocoa plantation management can help control CPB pests, as suggested by Silalahi [9]. In addition, Agung and Shahabuddin [12] found that polyculture was effective in reducing the percentage of cocoa beans damaged by CPB in Palopo District's Rahmat Village, while Mulyani and Iswahyudi [13] reported that farms treated with Integrated Pest Management (IPM) had lower CPB attack intensity in Aceh Tamiang District, Aceh Province. At the location of the efficacy study conducted in the cocoa plantations of the people of Central Maluku Regency, it was found that pest CPB attacks reached 80.7% [14].

The damage caused by the cocoa pod borer can result in significant yield losses and economic impact on cocoa farmers. Therefore, it is important to implement effective pest management strategies to control and prevent infestations by this pest. Integrated Pest Management (IPM) practices, such as the use of biological control agents and the implementation of cultural practices, can help to reduce the impact of CPB on cocoa production. Integrated Pest Management (IPM) emerged in Indonesia in the late 1980s in response to the environmental and social impacts of the Green Revolution. As a result, the United Nations Food and Agriculture Organization (FAO) and the Indonesian Government developed a cooperative program centered on Farmer Field Schools [15]. Integrated Pest Management (IPM) practices, such as the use of biological control agents and the implementation of cultural practices, have been effective in reducing the impact of CPB on cocoa production. Global market demand for cocoa will require higher standards of sustainability and other requirements of global environmental governance. Efforts made by the Indonesian government to increase cocoa production began in 2009 in Sulawesi to increase the competitiveness of the cocoa industry in the future with global market standards, through developing cocoa seeds tissue culture, increasing farmer capacity, and improving the quality of cocoa beans [16]. This chapter is a review that aims to present the implementation of IPM that has been carried out to control CPB in Indonesia and the role of other organizations as well as farmers' perceptions in the implementation of IPM in Indonesia. As the demand for cocoa is expected to increase in the coming years, it is essential to implement effective pest management strategies that are environmentally friendly and safe for farmers.

2. IPM implementation to CPB in Indonesia

2.1 Farmer Field School-IPM

The Farmer Field School-IPM (FFS-IPM) is one of the programs initiated by the Indonesian government to assist farmers in controlling pests in cocoa plants. This program is part of the Indonesian government's efforts to reduce the use of chemical pesticides that are harmful to human health and the environment.

Farmer Field Schools (FFS) were originally designed to improve the expertise and knowledge of farmers in various farming techniques (good cultivation techniques, the use of biological control agents, and the application of mechanical pest control), including the proper handling of pesticides (education to farmers on the dangers of using chemical pesticides and teaches them how to reduce their dependence on these pesticides). However, over time, the focus of the program shifted towards community organization, community planning, and Integrated Pest Management (IPM), leading to the development of Community IPM (CIPM). The principles of FFS have since been extended beyond rice to other crops, from IPM to plant breeding, and from technical domains to broader engagement with policy issues, advocacy, and local governance [15]. The Field School for Integrated Pest Management (IPM) was chosen as one of the methods to increase the knowledge and abilities of cocoa farmers in understanding cocoa pests.

The FFS-IPM aims to introduce IPM to a wider audience beyond the local level. This activity provides an opportunity for individual farmers or farming groups to develop their knowledge and skills through a 16-meeting training process at a location determined by the cocoa FFS-IPM participants. The cocoa FFS-IPM group participants will learn to analyze agroecosystems in their land and collaborate in creating plans to combat cocoa plant pests and disease infestations [17].

In the context of pest control in cocoa plants, FFS-IPM plays an important role in helping farmers increase their productivity and the quality of their harvest. By implementing Integrated Pest Management techniques, farmers can reduce losses caused by CPB, which can cause a decrease in cocoa production and quality. In the long term, FFS-IPM is expected to help reduce the use of chemical pesticides in the Indonesian agricultural sector and improve the welfare of farmers. This program can also help preserve the environment and strengthen sustainable agricultural systems in Indonesia.

2.2 The techniques of IPM implementation in Indonesia

The implementation of Integrated Pest Management (IPM) technology for controlling cocoa pod borer (CPB) in Indonesia involves several activities.

The training and education of farmers on the principles of Integrated Pest Management (IPM), which emphasizes the use of multiple methods to control pests, including cultural, biological, and chemical controls, is an active step taken. Farmers are actively being trained on the identification of cocoa pod borer (CPB) and the damage it causes to cocoa pods to determine the appropriate control measures to be applied. To monitor the population of CPB and prevent their spreading, farmers are being actively taught how to use pheromone traps. Pheromone traps are actively being used as devices that use synthetic sex pheromones to attract male CPB. By monitoring the number of trapped CPB, farmers can actively determine the severity of the infestation and the need for further control measures.

Cultural control methods, such as good agricultural practices (GAP), pruning, and sanitation, including removing infected pods plant parts, and debris from the field to prevent the pest's spread and minimize its habitat, are actively being implemented by farmers. GAP, which includes maintaining proper plant spacing, appropriate fertilizer application, and soil management, is actively being utilized to further reduce the population of CPB. Farmers are actively being instructed to intercrop with legumes such as soybeans or peanuts to improve soil fertility and reduce the incidence of CPB.

The farmers can use biological control agents, such as parasitic wasps, for controlling CPB. The natural enemies attack and kill the CPB larvae, thereby reducing their population in the field. Farmers are encouraged to release natural enemies, including parasitoids and predators, to use biological control methods. Parasitoids can be introduced to attack the CPB eggs and larvae, while predators like ants and spiders can help to control the adult CPB population. Farmers are also taught to use *Trichogramma* wasps to parasitize CPB eggs, leading to a reduction in the pest's population.

The use of pesticides is only recommended when other control measures have failed or when CPB populations exceed economic thresholds. Farmers are educated on the importance of following label instructions, wearing protective clothing, and using the correct application rates to reduce the risk of pesticide exposure and environmental damage. if the pest population is still high, farmers can apply chemical control measures as a last resort. However, the use of pesticides should be reduced to a minimum, and farmers should always follow the recommended dosage and safety procedures. Farmers have been trained on the proper use and handling of pesticides to minimize environmental contamination and the risk to human health.

The use of pesticides is only recommended when other control measures have failed or when CPB populations exceed economic thresholds. Farmers are educated on the importance of reducing the risk of pesticide exposure and environmental damage by following label instructions, wearing protective clothing, and using the correct application rates. If the pest population is still high, farmers can use chemical control measures as a last resort, but they are advised to reduce the use of pesticides to a minimum and always follow the recommended dosage and safety procedures, so to minimize environmental contamination and the risk to human health.

According to recent research conducted in East Aceh Regency, Indonesia, farms that have implemented Integrated Pest Management (IPM) practices have shown significantly lower levels of infestation by the cocoa pod borer (CPB) pests, as compared to farms that did not use any treatment [18]. In fact, the percentage and intensity of CPB attacks were found to be the lowest in farms that had implemented IPM practices. On the other hand, untreated farms showed the highest levels of infestation by CPB pests [13]. The use of IPM practices, along with cocoa pruning techniques and the use of black ants (*Dolichoderus thoracicus*) as natural enemies, can significantly reduce the incidence of CPB attacks in cocoa farms [18].

2.2.1 Monitoring and sex feromon

Monitoring the population of CPBs is also important in CPB control in Indonesia. Population monitoring can be done by installing pheromone traps in cocoa fields, especially in areas that have been infected with CPBs. When male CPBs smell the sex pheromones, they will be trapped in the trap. This can help farmers to monitor CPB density regularly, reduce the level of damage to cocoa pods and assess the efficacy of their control measures. Farmers need to monitor their cocoa trees regularly to detect CPB infestations early. They should also be able to identify the signs of CPB infestation, such as entry holes, frass, and damaged pods.

Pheromone traps are designed to attract and capture male CPB moths, thereby reducing the number of male moths available for mating and ultimately reducing the population of CPB. The pheromone traps contain synthetic chemicals that mimic the sex pheromones of female CPB, which attract male moths to the trap. This method can be effective in reducing CPB populations, but it must be used in conjunction with other control methods to achieve optimal results. The use of sex pheromones can be part of the strategy for controlling cocoa pod borer (CPB) in Indonesia. In CPB control, synthetic sex pheromones can be used to lure male CPBs into traps, thereby reducing the population of CPBs and reducing damage to cocoa pods. The use of sex pheromones to attract CPB at a density of 4 traps/ha can reduce yield losses by 67.7%. Use of sex pheromones for monitoring or mass trapping of CPB, as a component in IPM of CPB is promising, due to its nature for specific targets, environmentally friendly, effective, and economic values [19].

In using sex pheromones for CPB control, it is important to monitor the traps regularly and replace the pheromones every few weeks. This is important to ensure that the traps remain effective and do not attract other insects that are not related to CPBs. By combining monitoring techniques and the use of sex pheromones, farmers can optimize CPB control in their cocoa fields and reduce the economic losses caused by CPB infestations. Using sex pheromones obtained a total catch of insects in all blocks was 282 heads, the CPB attack category ranged from 4.38 to 16.398 with an attack intensity of 69% before application. After the application of sex pheromones, the average intensity of attacks dropped to 0.08% [20]. The pheromone trap was more useful for monitoring tools rather than for controlling CPB infestation [21]. Pheromone traps with a height of 1 m are the most effective CPB traps with a catch of 85 heads and the average population of trapped imago is 10.63 heads/month [22].

2.2.2 Cacao resistant clones to CPB infestation in Indonesia

It is important to develop cocoa varieties that are resistant to cocoa pod borer (CPB) in order to increase cocoa production in Indonesia. Several research studies have been conducted in Indonesia to produce cocoa varieties that are resistant to CPB. Some cocoa varieties are resistant or moderately resistant to CPB and have been successfully developed in Indonesia. Some come from breeding results and some come from local clones. ICCRI 07 and ICCRI 03 are clon cacao breeding results by Puslitkoka. Other clones come from Central Sulawesi (Sulawesi 02, MCC 01, MCC 02), South East Sulawesi (PT Ladongi, ARDACIAR 24, ARDACIAR 25, ARDACIAR 26), South Sulawesi (ARDACIAR 10), and North Sumatra (PABA/I/Pbrk, PABA/V/81 L/2, PABA/VIII/78B/3, PABA/VIII/78F/2, PABA/V/81 L/1, PABA/VIII/78B/1 (**Table 1**).

No	Clone name	Origin	Response to CPB infestation	Reference
1	ICCRI 07	Breeding by Puslitkoka 514	Resistant	[23]
2	ICCRI 03	Breeding by Puslitkoka	Moderate resistant	[24]
3	Sulawesi 02	Central Sulawesi	Resistant	[25, 26]
4	Sulawesi 03		Resistant	[23]
5	MCC 01	Central Sulawesi	Moderate resistant	[23]
6	MCC 02	Central Sulawesi	Resistant	[23, 27]
7	AP 70		Moderate resistant	[25]
8	PT Ladongi	South East Sulawesi	Resistant	[26]
9	ARDACIAR 10	South Sulawesi	Resistant	[28–30]
10	ARDACIAR 26	South East Sulawesi	Moderate resistant	[24]
11	ARDACIAR 24	South East Sulawesi	Resistant	[30]
12	ARDACIAR 25	South East Sulawesi	Resistant	[30]
13	PABA/I/Pbrk	North Sumatra	Moderate resistant	[13]
14	PABA/V/81 L/2	North Sumatra	Moderate resistant	[30]
15	PABA/VIII/78B/3	North Sumatra	Moderate resistant	[30]
16	PABA/VIII/78F/2	North Sumatra	Moderate resistant	[30]
17	PABA/V/81 L/1	North Sumatra	Resistant	[23, 30]
18	PABA/VIII/78B/1	North Sumatra	Resistant	[30]

Table 1.
Cocoa clones of resistant or moderate resistant to CPB in Indonesia.

Furthermore, several universities in Indonesia such as Jember University, Lampung University, and Bogor Agricultural Institute (IPB) have also conducted research to develop CPB-resistant varieties. Breeding for CPB resistance on cocoa in Indonesia was initiated by selecting resistant clones of the collected genotypes [31]. The process of selecting CPB-resistant genotypes takes time as the resistance has to be confirmed during several periods of harvest time to make sure the resistant expression would not be escaped the mechanism [24]. The collection performed their various resistance which were classified into five groups of resistance namely resistant, moderately resistant, moderately susceptible, susceptible, and highly susceptible to indicate the variability of CPB resistance [31]. Furthermore, breeding for CPB resistance will be designed by inter-crossing between selected-parental clones which perform differences in CPB resistance, yield potency, and genetic background [31].

Efforts have been made to explore the resistance of cocoa genotypes against CPB in various endemic areas in Indonesia, resulting in several promising resistant clones that have been used for breeding purposes [28]. The resistance of clones to CPB is influenced by both genetic and environmental factors. Using local resistant clones can effectively control CPB, promote efficiency and eco-friendliness, and improve productivity [27]. The selection of local resistant genotypes is an important strategy to control pests and diseases, enabling farmers to choose resistant clones without compromising productivity. Resistant clones that survive pest and disease outbreaks are selected, tested, and used to replace cocoa trees that have succumbed to the diseases [26, 32].

Understanding the resistance characteristics of cocoa pod borer (CPB) is crucial to identify selection criteria for choosing CPB-resistant clones. Different cocoa clones exhibit diverse responses to CPB attacks, indicating the complexity of the resistance mechanism [28, 29]. Morphological and anatomical characteristics of cocoa pods are important selection criteria for CPB resistance, such as fruit shape, skin thickness, and the presence of proteinase inhibitors [30]. However, factors other than pod hardness might be involved in resistance as suggested by the positive correlation of CPB incidence in ripe and immature pods, and the lower CPB incidence observed in some clones could be explained by pest non-preference. Several local selections, such as Aryadi 2 and Darwis 2, showed partial resistance to CPB, and two resistant clones, PT. Ladongi and Sulawesi 2, were identified with light levels of attack on beans [26, 32]. Selecting and planting resistant clones can effectively control CPB while maintaining optimal productivity.

However, the development of CPB-resistant varieties needs to be sustained. There are several challenges that need to be overcome in the development of CPB-resistant varieties, such as the sustainability and stability of resistance traits in the developed varieties. Therefore, research and development of CPB-resistant varieties need to continue in Indonesia to improve the quality and sustainability of cocoa production.

2.2.3 Cultural and mechanic control to CPB in Indonesia

There are several cultural methods that can be used to control cocoa pod borer (CPB) in Indonesia.

2.2.3.1 Sanitation

This method involves removing infected cocoa pods and debris from the field to eliminate the habitat of the CPB and reduce its spread. This is an important cultural control method for CPB management, as the pest can overwinter in fallen cocoa pods and debris. Farmers should remove and destroy any infected pods and plant parts as soon as they are observed. Good sanitation practices, such as removing fallen leaves and debris from the ground, can help to reduce the habitat of CPB and other pests. Fallen cocoa leaves and other plant debris can serve as a breeding ground for CPB, so removing them can help to prevent infestations.

2.2.3.2 Intercropping

Intercropping cocoa trees with other crops, such as legumes, can help to improve soil fertility and reduce the occurrence of CPB infestation. CPB may have difficulty adapting to new host plants, which can help to reduce their population. Additionally, intercropping can provide alternative sources of income for farmers and help to diversify their crops.

2.2.3.3 Crop rotation

Alternating cocoa crops with other crops can help to prevent the buildup of CPB populations in the soil, as the pest may have difficulty surviving in the absence of its preferred host. This method can be particularly effective when combined with other cultural control methods, such as sanitation and intercropping. Crop rotation can help to reduce the population of CPB by breaking the pest's life cycle. By rotating cocoa with other crops, such as legumes or vegetables, farmers can disrupt the pest's habitat and reduce the risk of infestation.

2.2.3.4 Trimming and pruning

Trimming and pruning cocoa trees can help to reduce the population of CPB by removing their hiding places. The pests often hide in the branches and crevices of cocoa trees, making it difficult to detect and control them. Regular pruning can also help to promote healthy tree growth and increase cocoa production. Pruning can help to manage CPB populations by removing branches or twigs that are infested with CPB. Pruning can also help to promote healthy tree growth and improve light penetration, which can reduce the risk of infestation.

2.2.3.5 Good agricultural practices (GAP)

Proper plant spacing, fertilization, and soil management can help to maintain healthy cocoa trees that are more resistant to CPB infestation. GAP includes maintaining proper plant spacing, appropriate fertilizer application, and soil management. Additionally, farmers should prune their cocoa trees to remove any infected or dead plant parts. The timing of fertilizer application can also play a role in CPB management. Overuse of nitrogen fertilizers can lead to the development of succulent growth on cocoa trees, which can attract CPB. By applying fertilizers at the right time and in the right amounts, farmers can promote healthy tree growth and reduce the risk of infestation.

2.2.3.6 Use of physical barriers

Physical barriers, such as nylon netting, can be used to prevent adult CPB moths from laying eggs on cocoa pods. This method can be particularly effective in small-scale cocoa farms, where the use of chemical pesticides may not be feasible. Wrapping protecting cocoa fruit from pest attacks, has been commonly practiced on various types of fruit. The effectiveness of packing cocoa pods with plastic bags to prevent pest attacks has been proven. If the cocoa pods are fired continuously for 30 months, the yield of dry beans increases. How to pack cocoa pods with plastic bags as follows. Choose young fruit that will be barked, 8–10 cm long, about 70–100 days old). The plastic bag used measures 30 × 15 cm with a thickness of 0.02 mm and both ends are open. The plastic bag is sheathed over the fruit and the mouth of the plastic bag is tied with a rubber band to the fruit stalk, the fruit is left covered until harvest. This method can prevent the laying of CPB pest eggs [33].

Application of cooling to cocoa plants can reduce PBK attack, *Promecotheca palmivora*, and the incidence of pencil wilt, thus potentially increasing the productivity of cocoa plants. In this study, it can be proven that bio kaolin sprayed on test trees can cover the surface of the fruit well. The closure of this layer is a physical obstacle for PBK pest insects to perch, puncture, and lay eggs on the surface of the fruit. From these results, it can be concluded that spraying bio kaolin increases the number of fruits free from PBK (13.79%). In addition, spraying with bio-kaolin either every one week or every two weeks also produces more PBK-free fruit compared to cloaking [34].

2.2.4 Biological control

The cocoa pod borer (CPB) is a major pest of cocoa crops in Indonesia and can cause significant economic losses. Biological control is a method of managing pests using natural enemies such as predators, parasitoids, and pathogens. Here are some of the kinds of biological control that can be used to control cocoa pod borer in Indonesia.

2.2.4.1 Parasitoids

Parasitoids are insects that lay their eggs inside the body of the host and eventually kill it. There are several species of parasitoids that attack the cocoa pod borer, including the parasitoid wasp, *Trichogrammatoidea bactrae*. This wasp is known to parasitize CPB eggs, which can reduce the population of the pest. Parasitoids are insects that are natural enemies of other insects, including pests like the cocoa pod Borer (CPB). They are a type of parasitic insect that lay their eggs inside the body of the host insect, and the parasitoid larvae develop inside the host, eventually killing it. Parasitoids are a type of biological control agent that can be used to manage pests like CPB in an environmentally friendly and sustainable manner. One of the advantages of using parasitoids for biological control is that they can be very effective at reducing pest populations. For example, studies have shown that parasitoid wasps like *T. bactrae* can parasitize up to 70–90% of CPB eggs in cocoa farms, leading to significant reductions in CPB populations. In addition, parasitoids are generally safe for the environment, as they do not leave behind any harmful residues or cause any collateral damage to non-target organisms.

However, there are some challenges associated with using parasitoids for biological control. One challenge is that parasitoids can be sensitive to environmental factors such as temperature, humidity, and pesticide use, which can affect their effectiveness. In addition, it can be difficult to ensure that the parasitoids are released at the right time and in the right quantities to have the desired impact on the pest population. Nonetheless, parasitoids are a promising option for controlling cocoa pod borer and other pests, and ongoing research is exploring ways to improve their efficacy and use in Integrated Pest Management programs.

2.2.4.2 Predators

Predators are insects that feed on other insects. There are several predator species that feed on the cocoa pod borer, such as ants, spiders, and certain beetles. One example of a predator that has been successfully used to control CPB is the ground beetle, *Lebia grandis*. Predators are natural enemies of pests like the cocoa pod borer (CPB) and can help to keep their populations under control. Insects that are predators feed on other insects, and there are several predator species that have been found to feed on the cocoa pod borer, including ants, spiders, and beetles. One example of a predator that has been successfully used to control CPB is the ground beetle, *Lebia grandis*. *Lebia grandis* is a small, black ground beetle that is commonly found in cocoa farms. The beetle is known to feed on CPB eggs, larvae, and pupae, making it an effective natural enemy of the pest. Studies have shown that *Lebia grandis* can significantly reduce CPB populations in cocoa farms when introduced at the right time and in sufficient numbers. One study conducted in Indonesia found that introducing *Lebia grandis* into a cocoa farm led to a significant reduction in CPB populations. The study involved releasing adult beetles into the cocoa farm at a rate of 3–4 beetles per cocoa tree. The researchers found that the beetles were able to establish populations in the cocoa farm and significantly reduce the number of CPB larvae and pupae.

Another study conducted in Indonesia found that certain species of ants can also be effective predators of the cocoa pod borer. The study found that the ant species, *Pheidole megacephala*, and *Oecophylla smaragdina*, were able to significantly reduce CPB populations when released into a cocoa farm. The ants were able to find and feed on CPB eggs and larvae, which led to a reduction in the number of adult CPB on the farm. These studies demonstrate that predators like ground beetles and ants can be effective natural enemies of the cocoa pod borer and can help to keep their populations under control. However, it's important to note that the effectiveness of predators can vary depending on the specific conditions in the cocoa farm and the surrounding environment. In addition, predators can be sensitive to environmental factors like temperature and humidity, which can affect their effectiveness. Nonetheless, predators are a promising option for controlling cocoa pod borer and other pests in an environmentally friendly and sustainable manner.

According to Robika et al. [35] that by increasing the number and colonies of black ants (*Dolichoderus thoracicus*) has a very real effect on reducing the intensity of PBK attacks (*Conopomorpha cramerella*) by increasing the number of PBK larvae. The results showed that the application of black ants on PBK larvae with 35 predators for 20 larvae.

2.2.4.3 Pathogens

Pathogens are microorganisms that cause disease in the pest. One of the most effective biological control agents for CPB is a fungus called *Beauveria bassiana*. This fungus infects and kills the CPB larvae. *B. bassiana* is a type of entomopathogenic fungus, which means that it is a fungus that can infect and kill insects. The fungus works by infecting the CPB larvae through contact with its spores. Once the spores penetrate the larvae's exoskeleton, the fungus grows inside the insect's body, eventually causing it to die. Studies have shown that *B. bassiana* can be an effective biological control agent for CPB when used correctly. In Indonesia, several studies have been conducted on the use of *B. bassiana* for the biological control of CPB.

One study conducted in Sulawesi, Indonesia, evaluated the efficacy of *B. bassiana* in controlling CPB in a cocoa farm. The study involved spraying the fungus on the cocoa trees at a rate of 1.5 g/l of water. The researchers found that the application of *B. bassiana* led to a significant reduction in the number of CPB larvae and pupae on the farm. The fungus was able to infect and kill the larvae, which led to a decrease in the number of adult CPB on the farm.

Another study conducted in South Sumatra, Indonesia, evaluated the effectiveness of *B. bassiana* combined with other biological control agents in controlling CPB. The study involved combining the use of *B. bassiana* with *Trichogramma* wasps and the parasitoid wasp, *Habrobracon hebetor*. The researchers found that the combination of biological control agents was effective in reducing CPB populations in the cocoa farm.

A study conducted in North Maluku, Indonesia, evaluated Bio-K, biopesticides with the active ingredient of *B. bassiana* to control CPB. The results show that biopesticides treatment significantly reduced pod damage by CPB with a decrease in the average number of cocoas in the amount of 7.69% and light intensity on CPB attacks was less than 20% [36].

2.2.5 Chemical control

Cocoa pod borer (CPB) is a major pest of cocoa in Indonesia, causing significant economic losses for farmers. Chemical control is one of the most commonly used methods to control CPB infestations. Here are the types of chemical control used in Indonesia:

2.2.5.1 Synthetic insecticides

Synthetic insecticides such as chlorpyrifos, cypermethrin, deltamethrin, and fenpropathrin are commonly used to control CPB. These insecticides can be applied as a foliar spray or soil drench. However, the repeated use of synthetic insecticides can lead to the development of resistance in CPB populations, which reduces their effectiveness over time. Synthetic insecticides such as chlorpyrifos, cypermethrin, deltamethrin, and fenpropathrin are commonly used to control CPB. These insecticides can be applied as a foliar spray or soil drench. However, the repeated use of synthetic insecticides can lead to the development of resistance in CPB populations, which reduces their effectiveness over time.

2.2.5.2 Botanical insecticides

Botanical insecticides such as neem oil and pyrethrum are derived from plants and are considered safer alternatives to synthetic insecticides. These insecticides have lower toxicity to non-target organisms and can be effective against CPB when applied at the right concentration and timing.

Botanical insecticides such as neem oil and pyrethrum have gained popularity as safer alternatives to synthetic insecticides for controlling cocoa pod borer (CPB) in Indonesia. Here are some additional details about these botanical insecticides: Neem oil: Neem oil is derived from the neem tree, and it contains several compounds that have insecticidal properties. Neem oil works by disrupting the growth and development of insect pests, including CPB. It is considered safe for humans and the environment, and it has low toxicity to non-target organisms. Neem oil can be applied as a foliar spray or soil drench, and it has been shown to be effective against CPB when applied at the right concentration and timing. Pyrethrum: Pyrethrum is derived from the flowers of certain species of chrysanthemum, and it contains compounds known as pyrethrins, which have insecticidal properties. Pyrethrum works by attacking the nervous system of insects, including CPB. It is considered safe for humans and the environment, and it has low toxicity to non-target organisms. Pyrethrum can be applied as a foliar spray or dust, and it has been shown to be effective against CPB when applied at the right concentration and timing.

It is important to note that while botanical insecticides are generally considered safer than synthetic insecticides, they can still have negative effects on non-target organisms if not used properly. Additionally, the effectiveness of botanical insecticides can vary depending on factors such as the concentration, timing, and application method. Therefore, it is important to use these insecticides in combination with other control methods as part of an Integrated Pest Management (IPM) approach to control CPB infestations.

2.2.5.3 Biopesticides

Biopesticides, such as *Bacillus thuringiensis* (Bt), are naturally occurring bacteria that can be used to control CPB. These bacteria produce a toxin that is toxic to CPB larvae, but harmless to humans and other animals. Bt can be applied to cocoa trees and pods as a preventive measure or as a treatment after an infestation has been detected. Study in West Sumatra, Indonesia: A study conducted in West Sumatra, Indonesia, controlled cocoa pod borer using a vegetable insecticide made from soursop leaves. The application of soursop leaf extract (*Anona muricata* L.) is able to control the best cocoa pod driving pest (*C. cramerella* Snellen) at a concentration of 100 g l^{-1} water which reduces the percentage of infected fruit to 10% with the lowest fruit damage intensity of 12.48%. The diameter of the largest fruit produced reached 18.15 cm with the smallest larval population of 0.75 tails, and the maximum seed dry weight reached 150.75 g [37].

3. Other organizations supported implementation of IPM in Indonesia

Upon analyzing the given text, it can be seen that various organizations are working to support the implementation of Integrated Pest Management (IPM) practices in cocoa plants in Indonesia. For instance, the Australian Centre for International Agricultural Research (ACIAR) funded a project to develop a locally applicable, farmer-participatory methodology for selecting and testing promising cocoa genotypes on farms in Southeast Sulawesi. In this trial, cocoa selections were propagated clonally and evaluated for two years for pod value, quality, and resistance to pests and diseases [32]. This demonstrates the efforts made to improve the resistance of cocoa plants to pests and diseases, which is crucial for sustainable cocoa production.

In addition to the ACIAR project, various NGOs and government agencies are also providing guidance and support for the implementation of IPM practices. The NGO of SWISS Contact, NGO Keumang, the Plantation Department, and the Counseling Agency at each sub-district in East Aceh Regency are some examples of organizations that provide guidance for implementing IPM practices [18]. These agencies not only provide guidance but also carry out regular monitoring to ensure that the IPM practices are being implemented effectively.

4. Perception of IPM implementation by cocoa farmers

Two separate studies conducted in Indonesia revealed the positive impact of adopting Integrated Pest Management (IPM) and the SL-PHT program on cocoa farming. In Sukoharjo 1 Village, Sukoharjo district, Pringsewu Regency, Lampung Province, cocoa farmers showed a positive perception towards the SL-PHT program, and the program's implementation led to an increase in cocoa plant productivity, income, and pest control [17]. The study also found a correlation between the farmers' level of experience, knowledge, social interaction, and their perception of the program's effectiveness. In the Ataku village, Andoolo Sub District, South Konawe District, Southeast Sulawesi Province, the implementation of IPM by cocoa farmers resulted in higher average income and greater production compared to those who did not adopt IPM. These studies suggest that IPM and the SL-PHT program can significantly improve cocoa farming outcomes, benefiting both the farmers and the industry as a whole.

The adoption of IPM practices among cocoa farmers can help to reduce the harmful effects of chemical pesticides on the environment and human health while promoting sustainable agriculture. Through the use of IPM techniques, farmers can improve crop yields and reduce the cost of pest management, which in turn can lead to increased incomes and improved livelihoods.

5. Conclusions

Cocoa pod borer (CPB) can cause significant damage to cocoa crops, resulting in economic losses for farmers. Integrated Pest Management (IPM) practices, such as biological control agents and cultural practices, can help reduce the impact of CPB on cocoa production. Preventive measures, including regular monitoring of cocoa pods, the use of resistant clones, and proper sanitation of farms, can also help. Implementing IPM practices can reduce the use of harmful chemical pesticides and promote sustainable agriculture. Organizations that support the implementation of IPM practices in Indonesia can help farmers improve their resistance to pests and reduce their reliance on pesticides. The adoption of sustainable agricultural practices, including IPM, can improve cocoa farming outcomes, and education and community engagement are essential for promoting their adoption and long-term sustainability. Research and development are needed to improve IPM strategies, including the development of new biological control agents and the optimization of cultural practices.

Author details

Araz Meilin*, Nurmili Yuliani, Nurhayati, Hermawati Cahyaningrum, Rein E. Senewe, Saidah, Herlina N. Salamba, Delima Napitupulu, Ismon Lenin, Risma Fira Suneth, Nur Imdah Minsyah, Handoko, Suparwoto, Endrizal, Busyra B. Saidi, Jumakir, Waluyo, Yardha, Muh. Asaad, Syafri Edi, Yustisia, Sigid Handoko, Nila Wardani and Julistia Bobihoe
National Research and Innovation Agency, Bogor, Indonesia

*Address all correspondence to: araz002@brin.go.id

References

[1] Direktorat Statistik Tanaman Pangan, Hortikultura, dan Perkebunan. Statistik Kakao Indonesia 2021. Jakarta: BPS RI; 2021. 76 p

[2] Alias A, Kelvin L. Cocoa pod borer. In: End MJ, Daymond AJ, Hadley P, editors. Technical Guidelines for the Safe Movement of Cacao Germplasm. Revised from the FAO/IPGRI Technical Guidelines No. 20. Rome, Italy: CacaoNet Biodiversity International; 2017

[3] Wardoyo S. The cocoa pod borer, a mayor Hidrance to cocoa development. Indonesia Agricultural Research Development of Journal. 1980;**2**:1-4

[4] Matlick BK. Machete Technology? What Small Farmers Need! In: 1st International Workshop on Sustainable Cocoa Growing; 30 March–2 April 1998. Panama City: Smithsonian Migratory Bird Center;

[5] Neilson J, Gusli S, Palinrungi R, Sau A. Integrating Smallholder Cocoa Farmers Within Quality-Driven Supply Chains: Learning from the PRIMA Project. Makassar: PENSA (Program for Eastern Indonesia SME Assistance); 2005

[6] Rubiyo S. Increased production and development of cocoa (*Theobroma cacao* L.) in Indonesia. Buletin Riset Tanaman Rempah dan Aneka Tanaman Industri. 2012;**3**(1):33-48

[7] Ruf F, Yaddang. Indonesia Cocoa Sector Assessment How to Help Indonesian Cocoa Farmers to RE-Invest in Cocoa? Synthesis for the World Bank. Montpellier, France: CIRAD, UMR Innovation; 2014

[8] Yudha PA, Dwipa I, Azwardi D. Exploration of cocoa pod pest and damage level in Lima Puluh Kota Regency, West Sumatra, Indonesia. Asian Journal of Advances in Agricultural Research. 2021;**17**:14-20. DOI: 10.9734/ajar/2021/v17i430202

[9] Silalahi AA. The effect of cocoa pod borer (*Conophomorpa Cramerella*) on cocoa production. Journal of Agriculture. 2022;**1**(01):1-3

[10] Indrayana K, Muhammad H. Study on environmentally friendly cocoa fruit borer (CFB) pest control study in Mamuju regency. J. Agrotan. 2017;**3**(1):102-114

[11] Azim SF, Kandowangko DS, Wanta NN. Damage of Cocoa Seeds by Fruit Bear Pest (*Conopomorpha cramerella Snellen*) on Cocoa Plantations in Muntoi and Solimandungan Villages. COCOS; 2016. Available from: https://ejournal.unsrat.ac.id/index.php/cocos/article/viewFile/11972/11561

[12] Agung A, Shahabuddin. The effect of share system on attacks of CPB, *Conophomorpha cramerella* (Gracillariidae: Lepidoptera) and cocoa production on cocoa plantation Lan in Rahmat Village, Palolo District. Agrotekbis. 2014;**2**(3):224-229

[13] Mulyani C, Iswahyudi I. Correlation of percentage and intensity of CPB attacks on production and loss of products due to various levels of cocoa plantation care. In: Proceedings of the E3S Web of Conferences; 24 September 2022; Bengkulu. Vol. 373, 07004. Indonesia: ISEPROLOCAL 2022; 2023. p. 9. DOI: 10.1051/e3sconf/202337307004

[14] Senewe RE, Wagiman FX, Wiryadiputra. Effectiveness of bioinsecticide bacillus thuringiensis

formulation against cocoa pod borer in field condition. Pelita Perkebunan. 2013;**29**(2):108-119. DOI: 10.22302/iccri.jur.pelitaperkebunan.v29i2.58

[15] Fakih M, Rahardjo T, Pimbert M. Community Integrated Pest Management in Indonesia. IIED, IDS, and REaD (member of INSIST); 2003. Available from: https://insistpress.com/katalog/community-integrated-pest-management-in-indonesia/ [Accessed: May 8, 2023]

[16] Cahya N, Atiyatul Maula K. Faktor-Faktor Yang Mempengaruhi Penawaran Dan Permintaan Bahan Pokok Di Indonesia. Transekonomika: Akuntansi, Bisnis dan Keuangan. 2021;**1**(4):311-320. DOI: 10.55047/transekonomika.v1i4.56

[17] Rendi R, Tubagus H, Helvi Y. Perception of farmers SL-PHT program for productivity and cocoa farming income. Journal IAA. 2014;**2**(3):301-308

[18] Mulyani C, Husni, Bakti D, Azhar. Measurement level of incursion cocoa pod borer (*Conopomorpha cramerella* snella) at community cocoa plantations in East Aceh regency, Aceh, Indonesia. IOP Conference Series: Earth and Environmental Science. 2021;**667**(1). DOI: 10.1088/1755-1315/667/1/012079

[19] Sulistyowati E. Effectiveness of sex pheromone in controlling cocoa pod borer, *Conopomorpha cramerella* (Snell.). Pelita Perkebunan. 2014;**30**(2):115-122

[20] Negara A. Tanggapan hama penggerek buah kakao *Conopomorpha cramerella* terhadap feromon seks dan intensitas serangannya di Kabupaten Parigi Mountong, Sulawesi Tengah. Prosiding Seminar Nasional Masyarakat Biodiversitas Indonesia. 2015;**1**:1654-1657. DOI: 10.13057/psnmbi/m010720

[21] Rahmawati D, Wagiman FX, Harjaka T, Putra NS. Detection of cocoa pod borer infestation using sex pheromone trap and its control by pos wrapping. Jurnal Perlindungan Tanaman Indonesia. 2017;**21**(1):30. DOI: 10.22146/jpti.22659

[22] Sahetapy B, Masuna ED, Darwanti D, Goo N. Pengaruh Ketinggian Perangkap Feromon Terhadap Penggerek Buah Kakao *Conopomorpha cramerella* Snell. (Lepidoptera: Gracillaridae). Agricultura. 2022;**32**(3):290. DOI: 10.24198/agrikultura.v32i3.35296

[23] Ministry of Agriculture. Hulu Hilir Kakao. Bogor: Pusat Perpustakaan dan Penyebaran Teknologi Pertanian; 2019. p. 104

[24] Susilo AW, Mawardi S, Witjaksono, Mangoendidjojo W. Pod characteristics of cocoa (*Theobroma cacao* L.) related to cocoa pod borer resistance. Pelita Perkebunan. 2015;**31**(1):1-13

[25] Martono B. Morphological Characteristics and Germplasm Activities in Cacao. Jakarta: IAARD Press; 2014

[26] Rubiyo, Dewi YA, Imran, Salim A, Baharudin C, Indrawanto, et al. Evaluation of yield and pest and disease resistance of cocoa clones in Kolaka District, Southeast Sulawesi, Indonesia. Biodiversitas. 2020;**21**(12):5698-5607. DOI: 10.13057/biodiv/d211215

[27] Syatrawati, Asmawati. The pest rate of cocoa pod borrer (*Conopomorpha cramerella* Snellen) on five local cocoa clones. AgroPlantae. 2015;**4**(1):25-29

[28] Susilo AW, Mangoendidjojo W, Witjaksono. Relationship between pod characteristics of some cocoa clones (*T. cacao* L.) and their resistance response to cocoa pod borer. Pelita Perkebunan. 2007;**23**(3):159-175

[29] Susilo AW, Mangoendidjojo W, Witjaksono, Mawardi S. The effect of pod age development of some cocoa clones to the expression of pod characteristics related to cocoa pod borer (CPB) resistance. Pelita Perkebunan. 2009;**25**(1):1-11

[30] Limbongan J. Karakteristik Morfologis dan Anatomis Klon Harapan Tahan Penggerek Buah Kakao Sebagai Sumber Bahan Tanam. Jurnal Litbang Pertanian. 2012;**31**(1):14-20

[31] Susilo AW, Zhang D, Motilal L. Assessing genetic diversity cocoa (*Theobroma cacao* L.) collection resistant to cocoa pod borer using simple sequence repeat markers. Pelita Perkebunan. 2013;**29**(1):1-9

[32] McMahon P, Iswanto A, Susilo AW, Sulistyowati E, Wahab A, Imron M, et al. On-farm selection for quality and resistance to Pest/diseases of cocoa in Sulawesi: (i) performance of selections against cocoa pod borer, *Conopomorpha cramerella*. International Journal of Pest Management. 2009;**55**(4):325-337

[33] Senewe RE, Wagiman FX. The position and wrapping of cocoa fruits to prevent Pest attack of Conopomorpha cramerella. Jurnal Budidaya Pertanian. 2010;**6**(1):21-24

[34] Irma K, Budiani A, Wahab A, Darmono d TW. Aplikasi biokaolin untuk perlindungan buah kakao dari serangan PBK, Helopeltis spp. dan Phytophthora palmivora. Menara Perkebunan. 2010;**78**(1):25-31

[35] Robika, Heri H, Zaedar A, Masese, dan Mihwan, Sataral. Pengendalian Hayati Hama Penggerek Buah Kakao (*Conopomorpha cramerella*) Menggunakan Semut Hitam (*Dolichoderus thoracicus*). Celebes Agricultural. 2020;**1**(1). p-ISSN: 2723-7974, e-ISSN: 2723-7966. Available from: https://ojs-untikaluwuk.ac.id/index.php/faperta

[36] Lala F, Cahyaningrum H, Hadiarto A, Brahmantiyo B. Implementation of biopesticides to control *Phytopthora palmivora* Butl. and *Conopomorpha cramerella* snell. On Cacao in South Halmahera. In: Proceedings of 5th International Conference on Food, Agriculture and Natural Resources (FANRes 2019): 17-19 September 2019, Ternate. Vol. 194. Atlantis Press: Advance in Engineering Research; 2020

[37] Rustam R, Hariyati R. Concentration test of soursop leaf (*Annona muricata* L) powder extract to control cocoa pod borer pests (*Conopomorpha cramerella* Snellen.) in cocoa (*Theobroma cacao* L) in Padang Pariaman. Agroekotek Journal. 2021;**13**(1):39-51

Chapter 9

Impact of Illegal Mining Activities on Cocoa Pollinator Abundance in Ghana

Sampson Addae, Sarah Acquah and Samuel Nyarko Essuman

Abstract

Cacao (*Theobroma cacao* L.) is mainly pollinated by Ceratopogonid midges (Forcipomyia spp.). Wild pollinators are important to both cocoa production and natural ecosystems, and are threatened by land-use change, intensive agricultural management, fragmentation from mining activities, and climate change. Despite the massive expansion of cocoa production in Ghana, it may now be of secondary economic importance to gold due to its increased environmental impact and the economic importance exercised by cocoa communities. However, very little attention has been paid to pollination management as a factor of production, as pollination is often not considered an important process for crop yield. The Ghana Cocoa Board takes a closer look at the impact of illegal mining on cocoa productivity and trusts farmers to destroy their farmland for illegal gold mining. In this chapter we briefly describe the cocoa sector, cocoa flower and pollinator biology and phenology as presented. What follows is an overview of the current environmental threats and social issues posed by unregulated mining affecting pollinator abundance and diversity in the context of Ghana. Finally, we examine ways to improve pollination and deforestation in Ghana's small cocoa sector.

Keywords: pollinators, illegal mining, habitat fragmentation, deforestation, cocoa, Ghana

1. Introduction

Pollination services are an essential element to maximize production in sustainable agriculture [1–3]. They are critical to food production and human livelihood, directly connecting wild ecosystems to agricultural production systems [4]. Wild pollinators are important in both agricultural and natural ecosystems and are threatened by land-use change [5], intensive agricultural management and pesticide use [6], pathogens [7] and climate change. The most important cocoa pollinators are midges. In cocoa, pollination occurs almost exclusively by ceratopogonid midges (order Diptera) of the genus Forcipomyia [8, 9]. The midges species that pollinate cocoa belong to the genus Forcipomyia. Most researchers studying cocoa pollination assume that Forcipomyia, which is common in cocoa plantations in Ghana, is Forcipomyia squamipennis I.

Demand for cocoa, the third largest trade commodity globally [10], is increasing in China, Russia, India, and Brazil by 2.5% per year [11], while production has declined by an average 1.5% annually for the past decade [12]. The three major cocoa producing countries are Ivory Coast, Ghana, and Indonesia. In West Africa, it is predicted that 89.5% of current cocoa growing areas will experience a decline in cocoa suitability by 2050 [13]. Businesses and governments are advocating sustainable cocoa initiatives to prevent deforestation and promote ecosystem services, biodiversity and sustainable livelihoods [14]. Cocoa production in Ghana is exclusively concentrated in the country's Ashanti, Brong, Ahafo, Central, Eastern, Western and Volta agroecological forest areas where the climatic conditions are ideal for cocoa production. The first cocoa farm was mostly established in south-east Ghana. Since then, the epicenter of production has gradually shifted west. In the early 1980s, the Ashanti and Brong Ahafo regions accounted for 35.5 and 18.5% of total production, respectively. Today, the western region alone supplies 56.5% of the total annual cocoa harvest. The cocoa industry generates about US$2 billion in foreign exchange annually and plays an important role in the national economy. The sector employs around 800,000 farming families in six of the country's ten regions [15]. Ghana is the second highest cocoa producing country in the world after Ivory Coast, with the sector producing approximately 970,000 tonnes of cocoa in 2017 [16].

Ghana's cocoa sector continues to make a significant contribution to the country's economy. In 2014 it accounted for 21.5% of the country's GDP and between 2009 and 2014 it averaged 25.6% [17], representing a large share (16.9%) of the agricultural sector's contribution to GDP. The environmental impacts of gold mining are wide-ranging and severe. A recent USAID study found that despite the massive expansion of cocoa production in the country, cocoa production may now be of secondary economic importance to gold due to its increased environmental impact and economic importance, even if mining is not practiced by all [15]. While Ghana has developed action plans to end deforestation in the cocoa sector and restore forest lands, actions to address the rapid expansion of artisanal and small-scale gold mining have so far been quite limited, jeopardizing potential gains in reducing deforestation in the cocoa sector.

In this chapter, we examine how habitat loss, fragmentation, deforestation, and climate effects affect the abundance and diversity of cocoa pollinators. We will first introduce the global cocoa sector and then focus on Ghana, the second largest cocoa producing country in the world. Next, we provide background on the biology and phenology of cocoa flowers and pollinating midges (Forcipomyia spp.), followed by a discussion of the impact of illegal mining on pollinator abundance and diversity (environmental and management) in Ghana. Finally, we discuss some options for easing restrictions on cocoa pollination.

2. The cocoa crop sector

2.1 Global cocoa production and economic value

Cocoa is a cash crop for many West African smallholder farmers. Although global annual cocoa production quantities are below those of other tropical crops such as sugar cane, rice, soybeans, oil palm, cassava or bananas, it is a unique crop as more than 90% of its production comes from smallholder farmers [18]. Cocoa production is estimated to be around 10 million hectares, which accounts for only 0.7% of the total global arable land but 7% of the global permanent land [19]. Therefore, cocoa cultivation and especially cocoa agroforestry systems play an important role in carbon sequestration and therefore have a significant mitigation potential [20]. Climate change, deforestation, diseases and low world market prices threaten global cocoa production. Global annual cocoa production has doubled in recent decades, reaching 3.6 million tonnes in 2009–2010, increasingly being concentrated in a handful of countries. Africa has established itself as a leading cocoa supplier over the past 10 years. According to the International Cocoa Organization, Africa's production has increased at an average annual rate of 2.7% since 2000. Farmers throughout the cocoa belt of West and Central Africa account for more than two-thirds of the world's cocoa production. World annual production reached 4.5 million tons in 2013 and growth came mainly from West Africa [18].

2.2 The cocoa sector in Ghana

Cacao is an important source of income for the provision of various public infrastructures and a profession mostly loved by farmers in Ghana and other countries. Cocoa dominates the agricultural sector and is an important source of income for approximately 800,000 farmers and many others involved in the trade, transport and processing of cocoa [21]. Ghana's cocoa sector has seen an impressive recovery in recent years. Ghana can boast of over 1,000,000 tons of cocoa in the 2010/2011 harvest year. Ghana is the second largest cocoa producer in the world [22, 23]. Together with Ivory Coast, it produced almost 52% of the world's total cocoa production in 2016. About 800,000 cocoa households produce 75 percent of all cocoa in Ghana [24]. The Ghanaian government generates most of its income from the export of cocoa [19]. Cocoa provides resources for poverty alleviation as well as food, money, employment and industrial raw materials [25]. The small size of farms in Ghana may be due to cocoa farm agreements where the farm is sometimes shared between the landowner and the caretaker [26]. The inheritance system in Ghana often leads to fragmentation of farmland when a farmer share his farm to several sons. Smaller farms result in decreased yields as small farms are a disincentive to invest, leading to reduced use of fertilizers and fungicides/pesticides [27]. The average cocoa acreage fell from 9.6 to 7.5 ha between 2008 and 2014 [28].

Ghana faces the problem of poor farm maintenance in terms of pest and disease control and low soil fertility. This is due to the low adoption of improved agricultural practices. For example, farmers only weed their farms twice a year on average, instead of the recommended four times. In addition, control of capsids and black pod disease occurs only twice a year instead of the recommended four or nine times a year [29]. Ghana has implemented the Hi-Tech program (fertilizer distribution via COCOBOD) to increase fertilizer use as fertilizer use in Ghana is low compared to Ivory Coast [30]. Timing of pesticide application is critical to maximize effectiveness in controlling mirids. The mirid population in West Africa begins to build up in July and peaks between August and September, while black pod occurrence increases from June and peaks in August and October. Therefore, it is recommended that cocoa plantations in Ghana are sprayed between July and September. Tetteh Quershie is the oldest variety still used in Ghana. It was named after the Ghanaian farmer who introduced cocoa to Ghana. The pods introduced by Tetteh Quarshie are of the Amelonado variety, which is a Forastero subspecies [31]. New introductions were made in 1944 from Forastero materials collected by F. J. Pound from the upper Amazon at the West African Cocoa Research Institute headquarters in Tafo, Ghana and Ibadan in Nigeria. Due to the earliness of these materials, they were widely used for replanting deforested plantations and by the late 1950s about 11 select species of the upper Amazon were being used to produce second and third generation Amazons, known as F3 Amazons or mixed Amazons, which were distributed to farmers. Several hybrid varieties have also been developed from these materials in Ghana, involving crosses with local Amelonado, Trinitario and some Criollo materials [32].

3. Cocoa pollination

The intensity of pollination and fruit set largely determine the cocoa yield [33]. However, very little attention has been paid to pollination management as a factor of production, as pollination is often not considered an important process for crop yield. Even when pollination is acknowledged, its management is often not considered. This is because most people believe that there are enough pollinators in the wild to carry out pollination without active human intervention.

3.1 Biology and phenology of cocoa fruit set

3.1.1 Biology

Cocoa flowers are hermaphrodite (male and female parts). Group of flowers (inflorescence) and pods are produced by the mature tree on the main stem and branches. This type of flowering is called cauliflory. The flowers are produced at the same defined areas on the tree and these swell with time to form flower cushions. Every cushion bears up to 50 flowers per flowering season. There are two flowering seasons per year, which thus yields 100 flowers per year. The pentamerous flower is about 15 mm in diameter. Flowers that are well pollinated develop to form fruits called pods. The main pod development stage is the cherelle or immature pod. A mature cocoa tree can produce over 10,000 flowers every year of which 1–50% are pollinated with 10–50 pods reaching maturity. The flowers first appear as small green or white buds at the flower cushion. The buds reach maturity and open within 28 days. The flower is small (0.5–1 cm), white, non-scented and borne on long pedicel or stalk. The female part consists of stigma, style and ovary which contains ovules (female sex cells) and the whole structure is surrounded by staminodes. The male part is made up of long stalk (filament) and anther with pollen grains (male sex cells) at the tip. The anther is hidden in a pocket- like structure called the petal sac or porch. Unlike most flowers where only the stigma is receptive to pollen, both the stigma and style are receptive to pollen. Most of the pollens are deposited on the style during pollination. Open flowers may remain on the tree for 48 to over 72 hours depending on the season, after which the unpollinated ones drop.

The sepals and petals of pollinated flowers start drying up giving it a brownish appearance. Greenish cherelles then emerges from the swollen ovary at the base of the female part. The presence of staminodes and petal porch prevent most insects from effectively pollinating the cocoa flower. Unlike most insect pollinated flowers, the color of the flower does not appear to attract its main pollinator, midges. However, there are purple colored lines called guide lines on the inner surface of the staminodes which guide them after landing on the flower. The midges upon landing on the staminode, move towards the base of the staminode as they feed along the guide lines and finally enters the porch which contains the pollen. There are hairs on the upper part of midges' thorax which are designed to pick pollen as they move within the porch.

Figure 1.
Stem of cocoa tree showing developed flower cushions with buds and flowers.

The insect then fly to another flower where the pollens are dropped on the stigma and style. This is made possible by brushing the thorax against the style as it moves to the base of the staminode (**Figure 1**).

3.1.2 Bud development

Flower bud development from meristem to receptive flower takes at least 20 to 30 days [34]. Prolonged dry (<125 mm per month) or cold (temperature < 23°C) periods inhibit flowering [35]. Flowering is optimal during rainy days with high relative humidity and moderate temperatures (100 mm per month, 70% RH, and 27°C). High solar radiation incidence is linked with increased flower abscission [34]. Pollen grains are only able to germinate on a receptive stigma [36]. The receptive period is at about 2–3 days after anthesis. Unsuccessful pollination leads to flower abscission. Reported flower abscission rates vary from 63% on the main trunk and 81% on the fan branches to over 90% for all flowers [34].

Anthesis starts around 2–4 pm. The process of sepal splitting continues overnight and finishes at around 4–6 am. Complete anthesis (flower fully open) is quickly followed by pollen release from the anthers (also between 4 and 6 am). Higher air temperature, as well as low air humidity, facilitates anther dehiscence [37]. However, pollen release is maximum between 8 am and 2 pm [38]. Styles and stigmas mature later than anthers, and have maximum receptivity around 12 am–2 pm. Maximum stigma and style receptivity does not concur with maximum anther dehiscence, thus limiting the possibility of self-pollination (**Figure 2**).

3.1.3 Cherelle wilt and pod maturation

Not all young fruits (cherelles) will grow to mature cocoa fruits, even after cocoa flowers are successfully pollinated and led to fruit set. Up to 80% of cherelles will shrivel, turn black, and become rapidly colonized by pathogens, while the pod remains on the tree. This so-called cherelle wilt is a physiological mechanism whereby

Figure 2.
Flower buds about to open and fully open flowers.

the fruits are naturally thinned to balance nutrient allocation in the tree. Cherelles can wilt up to day 100 after fruit set [39]. Poor soils and impeded photosynthesis result in increased cherelle wilting [35]. Leguminous shade trees, which supply nitrogen to the soil, can therefore lower cherelle wilt [40]. Wilting in an early stage saves energy that can be invested in the development of the remaining fruits [37].

It takes 5–6 months for pollinated flowers to become ripe pods. Different sizes or growth stages of flowers and pods can be found on the tree at any given time. Mature pods grow up to 30 cm long and 10 cm wide, and contain 20–60 beans (seeds) [37]. The cocoa fruit is an indehiscent drupe. During the first 40 days after fertilization, pod growth is slow. Afterward, growth accelerates. The first division of the zygote only takes place between day 40 and 50. Pod and ovule growth decrease from day 85 onwards, when embryos start to develop. On day 140, the embryo has completed its development and pod ripening starts [40].

3.1.4 Flower phenology

Cocoa bears fruits all year round, and the developmental stages start after pollination of the cocoa flowers occurs. However, only 1–5% of the flowers can successfully produce as a cocoa bud [39]. Some authors [41, 42] argued that the spatial arrangement of staminodes around the style of the cocoa flower affects pollination success and hence may limit fruit set. Others [43–45] also believed that cocoa flowers are nectarless and odorless. Young et al. [46] have demonstrated the presence of microscopic nectaries on the pedicels, sepals, and guide lines of the petals and staminodes that produce odor. These characteristics of the cocoa flower seem to make it unattractive to many potential pollinators, and therefore only insects that have evolved with the plant will successfully pollinate it. In most tropical countries, flowering occurs year-round. Flowering peaks are often preceded by increased temperature and rainfall, and occur at the onset of the rainy season, after which flower numbers gradually decline [47]. In West Africa, the major rainy season commences in April and climaxes in June, a period that is characterized by intense flowering (flowers on branches and trunks) [48].

In the minor rainy season (September–November), flowering intensity is lower (flowers on branches only). Few flowers are observed during the dry season (December–March) [49]. When pods are developing and this sink for assimilates is increasing, new flower production diminishes [26].

3.2 Biology and phenology of cocoa pollinators

3.2.1 Overview of cocoa pollinating species

Due to their viscosity, pollen grains form clumps and become too heavy to move independently [49]. Pollination of cocoa has the potential to overcome yield deficits in climate-resilient and sustainable production systems [12]. Pollination rates are generally poor for cacao and inconsistent throughout the year, so better pollination can increase yield. Experiments have been conducted in Ghana to increase the production of cocoa through hand pollination, which has been shown to increase fruit set, matured pods and the number of seeds per pod [49]. The Cocoa Research Institute of Ghana (CRIG) in Tafo used additional (artificial) hand pollination to increase yield and also breed new cocoa varieties [16, 49]. This aims to achieve maximum pollination, which is crucial for optimal yield in crop production, allowing for a bountiful harvest and increasing the export of cocoa beans, which will encourage farmers to increase production in the country [16].

Insect pollinators play an essential functional role in supporting ecological stability as well as food security worldwide [50]. Pollinating insects are vital to the world's food supply, pollinating more than 80% of the world's wild plant species [51]. The cocoa crop requires cross-pollination, which is mainly carried out by midges of the genera Ceratopogonidae and Cecidomyiidae [42]. Ceratopogonids are biting midges with a length of 14 mm [52]. It is believed that the females visit cacao flowers to feed on the protein-rich pollen grains necessary for egg maturation. Therefore, insufficient midges population leads to insufficient pollination and this deficiency has been reported as the main cause of low fruit set in some cocoa plantations [33, 53, 54]. In addition to midges, there are also wild bees Lasioglossum sp. and Hypotrigona sp., which visit cocoa flowers that bloom at tree canopy level to collect pollen. Lasioglossum in particular has been found to be effective at pollinating cacao flowers through its characteristic movement in and out of the petal porch. The role of Hypotrigona sp., which has been regularly identified on cocoa flowers in Ghana, has yet to be fully investigated. In general, more work needs to be done to fully understand the mechanisms of cocoa pollination globally and particularly in tropical Africa. In addition, other small Dipteran insects such as Cecidomyiidae (gall midges), Chironomidae (non-biting midges), Drosophilidae (fruit flies), Psychodidae (moth flies), and Sphaeroceridae (small dung flies) have been documented to visit cocoa flowers. Other insects such as aphids, coccids and cicadellids (Hemiptera), thrips (Thysanoptera) and ants (Hymenoptera) also occasionally visit cocoa flowers. However, their contribution to pollination is most likely very small. So far, pollen grains have been collected from insects other than Forcipomyia spp. not detected by microscopic observation. In some cases, observations suggest that cecidomyiids (in Cameroon) and drosophilids (in Ghana) may contribute to pollination to some extent [8]. Only Diptera and in particular the genus Forcipomyia (Fam. Ceratopogonidae) are morphologically capable of pollinating cocoa. Forcipomyia hosts the largest number of cocoa pollinators. Within this genus, the most commonly reported pollinators belong to the subgenera Euprojoannisia, Thyridomyia, and Forcipomyia [8].

These small midges, 2–3 mm long, prove to be excellent pollinators due to their frequency of visits and the massive deposition of pollen grains on the stigma [55].

Therefore, in order to maintain their population, pollinator abundance can be synchronized through environmental manipulation by providing suitable breeding sites in the cocoa field [56]. Providing breeding media in portable breeding containers can contribute to population increase in cocoa cultivation. Choices of growth media include cocoa pod husk (CPH), banana stumps; and combining cocoa pod shells with the availability of insect-infested pods in the cocoa field. Additional substrates provided must be easy to find in the cocoa field and will increase the population of Forcipomyia sp. increase [57]. However, the substrate must be replaced regularly as the moisture content decreases over time. Declining trends in pollinator populations can occur on various spatial and temporal scales [58], including reductions in floral resources. A lack of floral diversity, particularly in monoculture ecosystems, can limit the provision of resources needed by these midges [59].

3.2.2 Biology and phenology of Forcipomyia spp.

Adult midges can be found between the buttresses of large shade trees, in cracks of decayed old tree trunks, in hollow tree stumps and in piles of cocoa husks. Swarming occurs at any time of the day or in the late afternoon. Midges are present on cocoa plantations year-round, but the largest population occurs during the rainy season. Adult females lay eggs in batches of 40 to 90 on damp, rotting wood, cocoa husk, and other plant debris. The larvae hatch in 2–3 days and pupate after four molts when they are about 12 days old. The pupal stage lasts 2–3 days. Adult females require liquid plant food for survival and oviposition, although ovum maturation occurs independently of adult feeding or mating. Unfertilized eggs do not develop. The maximum life expectancy for either sex in captivity is eight days. Forcipomyia goes through at least 12 generations per year. Due to its abundance and continuous reproduction on cocoa plantations, Forcipomyia is probably the most important ceratopogonid cocoa pollinator in Ghana. According to [42], midges are attracted by the vertically aligned staminodes and use them to land. The fact that style pollination generally leads to greater fruit set than stigma pollination makes the ceratopogonid midges efficient pollination candidates [60].

The insect may then proceed into the petal hood along the purple colored guide lines where curved bristles on the thorax press against the anther, thereby picking up pollen grains. Both sexes visit cocoa flowers, but males appear to be more efficient pollinators. High numbers of Forcipomyia in the farm result in increased pod set. However, the number of pollinators depends on the availability of good breeding substrate such as cocoa pod husks, decaying plantain and/or banana stems in the farm. Ceratopogonid midge flights might cover long distances, but it is not known how far exactly [8]. Distance traveled during one foraging event, and consequently during which pollination is performed, can reach up to 50 m. It has been shown that there are 5–7 times more Forcipomyia species above the cocoa canopy than below the canopy [8]. Since wind speed above the canopy is higher than below, it can be expected that wind could play an important role in horizontal cocoa pollinator distribution over the cocoa field. Ceratopogonid pollinator populations can be abundant and exceed one million individuals per ha [9]. Moist environments favor ceratopogonid midge abundance. In fact, there is a positive correlation between soil moisture and ceratopogonid population levels [8]. Stable moist conditions are indispensable for successful development of eggs and larvae [61]. Pollinator populations thus increase with each rainy period, and decrease with the onset of a drier period [62].

4. Impact of illegal mining on pollinator abundance

4.1 Illegal mining activities and cocoa production in Ghana

The Ghanaian government generates most of its income from the export of cocoa [63]. Threats considered that lead to low cocoa yields include old aged trees (cocoa plantations over 30 years old), predominance of low-yielding traditional cultivars, smaller farm sizes due to fragmentation of land tenure agreements, illegal mining activities, and non-compliance with good agricultural practices [64, 65]. Artisanal and small-scale mining is understood to mean mining operations by individuals, groups, families or cooperatives with minimal or no mechanization, often carried out in the informal sector. The Ghanaian government recognized the potential of the sector for job creation and legalized artisanal mining in 1989 [66]. The Minerals and Mining Act 2006 (Act 703) further defines artisanal and small-scale mining as mining operations in an area corresponding to the prescribed number of blocks. In addition, small-scale mining was legalized as the exclusive domain of Ghanaians [66].

However, foreigners with sophisticated machinery found their way into the sector, accelerating the rate of extraction in the mining communities [66]. As a result, illegal small-scale mining activities increased after legalization aimed at regulating mining activities to protect the environment [66]. Small-scale illegal gold mining is referred to as Galamsey, which derives from the jumbling Ghanaian local jargon "gather-am-and-sell" [67]. Although illegal small-scale mining has drawn criticism from several quarters, the Ghana Cocoa Board takes a closer look at its impact on cocoa productivity and trusts farmers are destroying their farmland for illegal gold mining activities [68, 69]. In fact, the Ghana Cocoa Board has been one of the major complainers about the Galamsey threat [69–72], as Galamsey creates factors that discourage or adversely affect cocoa cultivation. First, Galamsey presents pull factors as a more attractive investment. Some people are drawn to informal mining because they believe mining offers them a get-rich-quick opportunity [73]. The farmers who have their cocoa farms close to the mining areas observe early dropping of immature pods, wilting and yellowing of leaves because the galamsey activities deplete the topsoil which supports the healthy growth of plants [74, 75]. The damaging environmental impacts associated with the unregulated mining activities include effluent damping, unrehabilitated excavations, improperly stored waste, dust emissions, deforestation, acid mine, river siltation and the release of chemicals such as cyanide and mercury [76, 77] asserted that between 1 and 20 hectares of cocoa lands are been taken over by galamsey activities in numerous Ghanaian cocoa-producing districts annually. According to GSS (Ghana Statistical Service) (2018), the GDP contribution of cocoa decreased from 3.6% in 2011 to 1.8% in 2017. The rife mining activities in the nation have been partly blamed for the reduction in cocoa production and economic contribution.

4.1.1 Landscape degradation

Habitat destruction from land-use changes such as habitat loss, fragmentation, deforestation and conversion of natural habitats to cropland is the most important driver of biodiversity loss in terrestrial ecosystems [78]. From an ecological perspective, changes in land cover involve shifts in land cover composition and variations in their spatial arrangement [79], which directly affect the composition of biological communities and pollinator-flowering plant relationships [80, 81]. Species that survive in such environments need to adapt to changing habitats and periodic disturbances. The integration of shade trees into cocoa agroforests can bring numerous economic and environmental benefits [82, 83], such as Increased Diptera visitation rates with increased canopy closure found in Indonesia [84].

Therefore, the long-term conversion of forests to mining sites in Ghana could result in agroecological disadvantages such as forest degradation, biodiversity loss, soil quality disruption associated with low yields and food insecurity [83–85]. A recent report estimates that the annual cost to the global economy of habitat loss will reach $10 trillion by 2050, making ongoing biodiversity loss as much an economic crisis as an environmental crisis. For more sustainable agriculture, three complementary strategies are envisaged that address several key drivers of pollinator decline: ecological intensification, strengthening existing diverse farming systems, and investing in ecological infrastructure [83]. These three strategies simultaneously address several key drivers of pollinator decline by mitigating the impacts of land-use change from illegal mining, pesticide use and climate change. Protecting large areas of semi-natural or natural habitat (tens of hectares or more) helps conserve pollinator habitats at a regional or national scale [84].

4.1.2 Habitat fragmentation

Habitat destruction, fragmentation and degradation, combined with conventional intensive farming practices, often result in reduced or altered pollinator food and nesting resources. It is well known that habitat destruction can reduce the population size, composition and species richness of pollinator communities [85, 86], and thereby affecting evolutionary processes at the species level. Significant declines have already been observed for some pollinator groups (e.g. Hymenoptera, Lepidoptera), which may be due in part to a history of habitat conversion [87] as well as the loss of certain habitat elements such as nesting or foraging sites [81, 88]. Differences in ecological and morphological traits (feeding adaptation, mobility, body size, behavior) can affect the response of pollinator species to changing environments and their ability to survive in poor quality environments [86]. Pollinator species that are more specialized for habitat or food requirements tend to be more vulnerable to land cover changes that alter the availability of food or nesting resources [87], leading to homogenization of pollinator communities dominated by common generalist species [89]. Gene flow has a major impact on genetic variation within populations, as it offsets the detrimental effects of genetic drift, determines effective population size, and has important implications for the management and conservation of genetic resources [82]. Since pollen movement is a key component of gene flow, density effects can be expected to alter genetic structure and, especially in small populations, increase the likelihood of extinction [86]. Therefore, given that tropical forests experience high rates of deforestation [90], knowledge of gene flow is fundamental to understanding reproductive success and management of tropical tree species.

4.1.3 Deforestation

Crop loss occur when the Galamsey operations are done directly on the farm. Cocoa crops are being destroyed by large machines such as bulldozers used to clear land on Galamsey farms. Loss of crop yield and income usually occurs when the Galamsey operators forcibly take part of the farmland from the farmers. This has several negative impacts on the environment, including water and air pollution, deforestation and land degradation. Tropical forests are the most biodiverse ecosystems and species-rich habitats in the world, but are significantly threatened by widespread ongoing deforestation [84]. Primary forests are of vital importance for the conservation of biological diversity due to their unprecedented diversity of species and habitats [82, 91]. Deforestation of primary and secondary forests can result in biodiversity loss and forest fragmentation, reducing previously uninterrupted habitat to smaller fragments [87]. This can result in forest-dependent species being isolated in small patches of forest that are not large enough to support a healthy population. However, secondary forests play a key role in providing habitat for a wide variety of species and in establishing connections between primary forest areas. Protection and restoration of primary and secondary forests are critical to improving tropical forest health [92]. Beyond habitat loss, land-use change can lead to deterioration in habitat quality, known as habitat degradation. In these cases, the species are able to survive, but their populations may decline [89]. For example, a recent study suggested that agricultural expansion has reduced the richness and composition of pollinators of bees and wasps in the UK [83].

5. Conclusions

Since cocoa production is largely dependent on pollination by insects, any threat to pollinators will negatively impact cocoa production. There is evidence that cocoa pollination is currently below optimal levels in Ghana and that increasing pollinator populations in cocoa fields could increase cocoa production [85]. This was demonstrated by the Cocoa Research Institute of Ghana (CRIG) in Tafo using additional (artificial) hand pollination to increase yield and also to breed new cocoa varieties [16]. It is noted that illegal small-scale mining (Galamsey) has actually been the major factor affecting cocoa production due to land degradation, water and air pollution, diversion of water bodies, damage to farms and farmhouses, etc. The findings from this suggest that Ceratopogonidae midges alone were probably too rare to act as the main or even sole pollinator of cocoa in Ghana, since the relative abundance observed on cocoa flowers in Ghana are low due to the operations of illegal mining activities done right in the cocoa farms [84].

The study identified the need for protection of extensive natural forest areas to protect the genetic identity of wild cacao pollinators in Ghana and, in addition, to promote genetic exchange between wild populations to maintain genetic variability of viable populations [22]. Evidence has shown that areas of secondary forest surrounding cocoa farms may provide pollinator resources similar to natural forests, and that deforestation and habitat fragmentation reduce the population size, composition and species richness of pollinator communities in cocoa farms to increase production [93]. Different pollinator species respond differently to changing environmental conditions caused by illegal small-scale mining due to their physiological, behavioral or other mechanisms [85].

6. Recommendation

6.1 Ecological intensification

Management of nature's ecological functions is needed to improve cocoa production and livelihoods while minimizing environmental damage.

6.2 Strengthening existing diverse farming systems

This includes the management of systems such as forest gardens and agroforestry to promote pollinators through scientifically and local knowledge validated practices, for example, Crop rotation, cultivation of sunflowers and edge crops to provide alternative food resources to increase pollinator abundance and diversity necessary for cocoa production in Ghana.

6.3 Policies and laws concerning Galamsey

Existing legislation should be strengthened by involving farmers in stakeholder decision-making on Galamsey, so that offenders are punished and others are deterred. To achieve this, the government should provide COCOBOD with structures that enable it to effectively and efficiently impose sanctions against illegal small-scale mining.

6.4 Strategies to conserve pollinators and biodiversity

Mining activities should be structured to strike a balance between economically driven extraction of mineral resources and the strategies needed to conserve natural resources and maintain ecosystem integrity and species viability.

Acknowledgements

The authors acknowledge the support from the Africa Regional Postgraduate program in insect science, co-authors, DAAD support and the cocoabod of Ghana.

Conflict of interest

The authors have no conflict of interest.

Author details

Sampson Addae*, Sarah Acquah and Samuel Nyarko Essuman
University of Ghana, Legon Campus, Accra, Ghana

*Address all correspondence to: trustaddae@gmail.com

References

[1] Kevan PG, Clark EA, Thomas VG. Insect pollinators and sustainable agriculture. American Journal of Alternative Agriculture. 1990;**5**:13-22

[2] Knapp JL, Bartlett LJ, Osborne JL. Re-evaluating strategies for pollinator-dependent crops: How useful is parthenocarpy? Journal of Applied Ecology. 2017;**54**:1171-1179

[3] Kovacs-Hostyanszki A, Espindola A, Vanbergen AJ, et al. Ecological intensification to mitigate impacts of conventional intensive land use on pollinators and pollination. Ecology Letters. 2017;**20**:673-689

[4] Lázaro A, Alomar D. Landscape heterogeneity increases the spatial stability of pollination services to almond trees through the stability of pollinator visits. Agriculture, Ecosystems & Environment. 2019;**279**:149-155

[5] Guzman A, Chase M, Kremen C. On-farm diversification in an agriculturally-dominated landscape positively influences specialist pollinators. Frontiers in Sustainable Food Systems. 2019;**3**:87

[6] Gaffney J, Bing J, Byrne PF, et al. Science-based intensive agriculture: Sustainability, food security, and the role of technology. Global Food Security. 2019;**23**:236-244

[7] Graystock P, Blane EJ, McFrederick QS, et al. Do managed bees drive parasite spread and emergence in wild bees? International Journal for Parasitology: Parasites and Wildlife. 2016;**5**:64-75

[8] Winder JA. Field observations on Ceratopogonidae and other Diptera: Nematocera associated with cocoa flowers in Brazil. Bulletin of Entomological Research. 1977;**67**(1):57-63

[9] Winder JA. Cocoa flower Diptera; their identity, pollinating activity and breeding sites. PANS. 1978;**24**(1):5-18

[10] Donald PF. Biodiversity impacts of some agricultural commodity production systems. Conservation Biology. 2004;**18**:17-38

[11] Squicciarini MP, Swinnen JF. The Economics of Chocolate. USA: Oxford University Press; 2016

[12] FAOSTAT. 2018. Available from: http://www.fao.org/faostat/en/#data/QC

[13] Läderach P, Martinez-Valle A, Schroth G, Castro N. Predicting the future climatic suitability for cocoa farming of the world's leading producer countries. Ghana and Côte d'Ivoire. Climatic Change. 2013;**119**:841-854

[14] IDH. Cocoa. 2018. Available from: https://www.idhsustainabletrade.com/sectors/cocoa/

[15] McDermott CL, Acheampong E, Arora-Jonsson S, Asare R, de Jong W, Hirons M, Setyowati A. SDG 16: Peace, Justice and Strong Institutions–A Political Ecology Perspective. Sustainable Development Goals: Their Impacts on Forests and People. United Kingdom: Cambridge University Press; 2019. ISBN: 978-1-108-48699-6, eISBN: 9781108765015

[16] COCOBOD. Environmental and Social Management Plan 3.1. Accra Ghana: Ghana Cocoa Board; 2018. pp. 4-5

[17] World Cocoa Foundation WCF. Cocoa & Forests Initiative 2017. Available

from: https://www.worldcocoafoundation.org/cocoa-forests-initiative/.

[18] ICCO. In: International Cocoa Organization, editor. Annual Report. Vol. 76. Abidjan, Côte d'Ivoire: ICCO; 2017

[19] Anaafo D. In pursuit of pro-poor development in Ghana: Is accessible, affordable and reliable energy supply the missing link? In: The Political Economy of Energy in Sub-Saharan Africa. Accra Ghana: Routledge (ISSER); 2017. pp. 30-46

[20] Silatsa FBT, Yemefack M, Ewane-Nonga N, Kemga A, Hanna R. Modeling carbon stock dynamics under fallow and cocoa agroforest systems in the shifting agricultural landscape of Central Cameroon. Agroforestry Systems. 2017;**91**(5):993-1006

[21] Ghana Cocoa Board (COCOBOD). Supply Chain Risk Assessment Report. Accra Ghana: COCOBOD; 2001. p. 3-7

[22] Boateng DO, Nana F, Codjoe Y, Ofori J. Impact of illegal small scale mining (Galamsey) on cocoa production in Atiwa district of Ghana. International Journal of Advance Agricultural Research. 2014;**2**:89-99

[23] Afele JT, Dawoe E, Abunyewa AA, Afari-Sefa V, Asare R. Carbon storage in cocoa growing systems across different agroecological zones in Ghana. Pelita Perkebunan (a Coffee and Cocoa Research Journal). 2021;**37**(1):32-49

[24] Maguire-Rajpaul VA, Khatun K, Hirons MA. Agricultural information's impact on the adaptive capacity of Ghana's smallholder cocoa farmers. Frontiers in Sustainable Food Systems. 2020;**4**:28

[25] Peprah K. Sustainability of cocoa farmers' livelihoods: A case study of Asunafo District, Ghana. Sustainable Production and Consumption. 2015;**4**:2-15

[26] Barrientos SW, Asenso Akyere K. Mapping Sustainable Production in Ghanaian Cocoa, Report to Cadbury. Ghana: Institute of Development Studies & University of Ghana; 2012

[27] Steijn CPA. Towards Sustainable Cocoa Production: A Mixed Method Assessment of the Influence of Local Governance Modes on the Farm Level Impact of Private Cocoa Certification Standards in Ghana [Thesis]. The Netherlands: Utrecht University; 2016

[28] Tulane University. Survey Research on Child Labor in West African Cocoa Producing Regions, 2013/14. USA: School of Public Health and Tropical Medicine, Tulane University; 2015

[29] Aneani F, Anchirinah VM, Asamoah M, Owusu-Ansah F. Analysis of economic efficiency in cocoa production in Ghana. African Journal of Food, Agriculture, Nutrition and Development. 2011;**11**(1);123-137

[30] Baah F, Anchirinah V. Looking for convergence: Stakeholders' perceptions of cocoa extension constraints in Ghana. Journal of Science and Technology (Ghana). 2010;**30**(3):1-13

[31] Kuapa Kokoo. Ghana and cocoa. 2017. Available from: https://www.kuapakokoo.com/about/ghana-and-cocoa [Accessed: November 29, 2017].

[32] Addae S. The Cocoa Certification Program and its Effect on Sustainable Cocoa Production in Ghana: A Case Study in Upper Denkyira West District [Thesis]. KUMASI: KWAME NKRUMAH UNIVERSITY OF SCIENCE AND TECHNOLOGY; 2014

[33] Young AM. Habitat differences in cocoa tree flowering, fruit-set, and pollinator availability in Costa Rica. Journal of Tropical Ecology. 1986;**2**:163-186

[34] Mohanaramya M. Studies on Floral Biology, Hybridization and Performance of Elite Clones of Cocoa (Theobroma cacao L.) in Different Locations of Tamil Nadu. India: Horticultural College & Research Institute. Coimbatore: Tamil Nadu Agricultural University; 2013

[35] Asomaning E, Kwakwa R, Hutcheon W. Physiological studies on an Amazon shade and fertilizer trial at the cocoa research institute, Ghana. Ghana Journal of Agricultural Science (Ghana). 1971;**4**(1):47-64

[36] Sanzol J, Rallo P, Mar H. Asynchronous development of stigmatic receptivity in the pear (Pyrus Communis; Rosaceae) flower. American Journal of Botany. 2003;**90**(1):78-84

[37] González MV, Coque M, Herrero M. Papillar integrity as an indicator of stigmatic receptivity in kiwifruit (Actinidia Deliciosa). Journal of Experimental Botany. 1995;**46**(2):263-269

[38] Schroth G, Lehmann J, Rodrigues MRL, Barros E, Macêdo JLV. Plant-soil interactions in multistrata agroforestry in the humid Tropicsa. Agroforestry Systems. 2001;**53**(2):85-102

[39] McKelvie AD. Cherelle wilt of cacao I. pod development and its relation to wilt. Journal of Experimental Botany. 1956;**7**(2):252-263

[40] Bos MM, Steffan-Dewenter I, Tscharntke T. Shade tree management affects fruit abortion, insect pests and pathogens of cacao. Agriculture, Ecosystems & Environment. 2007;**120**(2):201-205

[41] Glendinning DR. Natural pollination of cocoa. New Phytologist. 1972;**71**(4):719-729

[42] Kaufmann T. Studies on the ecology and biology of a cocoa pollinator, Forcipomyia squamipennis I. and M. (Diptera, Ceratopogonidae), in Ghana. Bulleting of Entomological Research. 1975;**65**:263-268

[43] Kaufmann T. Preliminary observations on a cecidomyiid midge and its role as a cocoa pollinator in Ghana. Ghana Journal of Agricultural Science. 1973;**6**:193-198

[44] Murray DB. In: Wood GAR, editor. The Botany of Cocoa in Cocoa. 3rd ed. London, UK: Longman; 1975. pp. 7-18

[45] McGregor SE. Insect Pollination of Cultivated Crop Plants. USA: US Department of Agriculture Handbook; 1976

[46] Young AM, Schaller M, Strand M. Floral nectaries and trichomes in relation to pollination in some species of Theobroma and Herrania (Sterculiaceae). American Journal of Botany. 1984;**71**(4):466-480

[47] Omolaja S, Aikpokpodion P, Oyedeji S, Vwioko D. Rainfall and temperature effects on flowering and pollen productions in cocoa. African Crop Science Journal. 2009;**17**(1):41-48

[48] Goenaga R, Guiltinan M, Maximova S, Seguine E, Irizarry H. Yield performance and bean quality traits of cacao propagated by grafting and somatic embryo-derived cuttings. HortScience. 2015;**50**(3):358-362

[49] Adjaloo MK, Oduro W, BKB B. Floral phenology of upper Amazon cocoa trees: Implications for reproduction and productivity of cocoa. ISRN Agronomy. 2012;**2012**:461674

[50] Crenna E, Sala S, Polce C, Collina E. Pollinators in life cycle assessment: Towards a framework for

impact assessment. Journal of Cleaner Production. 2017;**140**:525-536

[51] Klein AM, Vaissiere BE, Cane JH, Steffan-Dewenter I, Cunningham SA, Kremen C, et al. Importance of pollinators in changing landscapes for world crops. Proceedings of the Royal Society of London B: Biological Sciences. 2007;**274**:303-313

[52] Lane RP, Crosskey RW. Medical Insects and Arachnids. Netherlands: Springer; 2012

[53] Cilas C, Mississo E, Dufour B, Amefia YK. Etude des variations de la pollinisation naturelle du cacaoyer au Togo. Influence sur le remplissage en graines des cabosses - Study of the variations in natural pollination of cocoa tree in Togo. Influence on the pod seed filling rate Cafe, Cacao, the (Paris). 1987;**31**(4):283-290

[54] Ghani Ibrahim, A. (1988). ActasEffects of insect pollinators on fruit set of cocoa flowers in MalasiaProceedings. In 10. International Cocoa Research Conference17-23 May 1987Santo Domingo (R. Dominicana) (No. 633.74063 I61p 1987). Cocoa Producers' Alliance, Lagos (Nigeria)

[55] Adjaloo MK, Oduro W. Insect assemblage and the pollination system in cocoa ecosystems. Journal of Applied Biosciences. 2013;**62**:4582-4594

[56] Azhar I, Wahi SM. Pollination ecology of cocoa in Malaysia: Identification of taxonomic composition and breeding sites, ecology and pollinating activities, and seasonal abundance. In: Chew PS, Pushparajah EP, editors. Cocoa and Coconut: Progress and Outlook. Kuala Lumpur: Universiti Malaya; 1984. pp. 1-15

[57] Saripah B, Alias A. The use of transferable breeding container for population enrichment of pollinator, Forcipomyia spp. In the cocoa ecosystem. Journal of Advanced Agricultural Technologies. 2018;**5**(3):245-251

[58] Novais SMA, Nunes CA, Santos NBD, Amico AR, Fernandes GWQ, Braga RF, et al. Effects of a possible pollinator crisis on food crop production in Brazil. PLoS One. 2016;**13**(5):e0167292

[59] Blaauw BR, Isaacs R. Flower plantings increase wild bee abundance and the pollination services provided to a pollination-dependent crop. Journal of Applied Ecology. 2014;**51**(4):890-898

[60] Posnette AF, Entwistle HM. The pollination of cocoa flowers. Rpt. In: Cocoa Conference. London, UK: Grosvenor House; 1958. pp. 66-69. Plant Breeding, vol. 28,no. 4,pp. 4550, 1958

[61] Soria SDJ, Abreu JM. Dinamica Populacional De Forcipomyia Spp. (Diptera, Ceratopo gonidae) Na Bahia, Brasil; 1: Flutuacao Estacional Dos Polinizadores Do Cacaueiro Relacionada Com Chuva E Balanco De Água (Thornthwaite). Revista Theobroma (Brasil). 1976;**6**(2):47-54

[62] Wood GAR, Lass RA. Cocoa. 4th ed. London & New York: Longman; 1985

[63] World Bank Group. Third Ghana Economic Update: Agriculture as an Engine of Growth and Jobs Creation. Washington, D.C: World Bank; 2018. pp. 1-63

[64] Cobbina J. Technical Efficiency of Cocoa Production in Ghana, a Case Study of Upper Denkyira East Municipality [Thesis]. Ghana: Kwame Nkrumah University of Science and Technology; 2014

[65] Anthony E, Agbongiarhuoyi O, Taiwo FB, Solomon and O. Joseph

Olumide. Perception of cocoa farmers to voluntary standard certification: An implication on cocoa transformation in Nigeria. IOSR Journal of Agriculture and Veterinary Science. 2014;**7**:17-20

[66] Ministry of Lands and Natural Resources. Project Appraisal & Implementation Document for the Multilateral Integrated Project. Accra: JMK Consulting, LLc; 2017

[67] Duodu C. How galamsey is ruining rural Ghana. New African (531). 2013. p. 38. Available from: http://search.ebscohost.com/login.aspx?direct=true&db=f5h&AN=89907230&site=ehost-live [Accessed: May 20, 2017]

[68] Yeboah S. 'Crops' or 'Carats'? Gold mining and cocoa production in Ghana. 2014. Available from: https://www.linkedin.com/pulse/crops-carats-gold-mining-cocoa-stephen/

[69] Ghana Cocoa Board. Stop destroying cocoa farms for gold. 2017. Available from: https://cocobod.gh/news_details/id/116/STOP%20DESTROYING%20COCO%20FAR MS%20FOR%20GOLD%20%E2%80%93%20COCOBOD%20WARNS

[70] Addah R. Assessing the timpacts of galamsey operations on the socioeconomic activities of Ghanaian rural mining communities: A case study of Ayanfuri (Doctoral dissertation) Ashesi University College Accra Ghana. 2014

[71] Samuel A, Oladejo N, Adetunde I. The impact and effect of illegal mining (galamsey) towards the socio-economic development of mining communities: A case study of Kenyasi in the Brong Ahafo region. International Journal of Modern Social Sciences. 2012;**1**(1):38-55

[72] Burrows E, Bird L. Gold, guns and China: Ghana's fight to end galamsey. 2017. Available from: http://www.myjoyonline.com/opinion/2017/June-13th/gold-guns-and-china-ghanas-fight-to-end-galamsey.php [Accessed: October 23, 2017]

[73] Hilson G. Poverty traps in small-scale mining communities: The case of sub-Saharan Africa. Canadian Journal of Development Studies. 2012;**33**(2):180-189. DOI: 10.1080/02255189.2012.687352

[74] Aneani F, Adu-Acheampong R, Sakyi-Dawson O. Exploring opportunities for enhancing innovation in agriculture: The case of cocoa (Theobroma cacao L.) production in Ghana. Sustainable Agriculture Research. 2017;**7**(1):33. DOI: 10.5539/sar.v7n1p33

[75] Onumah J, Leeuwis C, Boamah PO, Salifu T. Operations and frames of illegal small-scale mining in Ghana. International Journal of Agriculture Innovations and Research. 2013;**2**:307-312

[76] Eftimie A, Heller K, Strongman J, Hinton J, Lahiri-Dutt K, Mutemeri N. Gender Dimensions of Artisanal and Small-Scale Mining: A Rapid Assessment Toolkit. Washington DC: World Bank Group; 2012

[77] Boafo J, Paalo SA, Dotsey S. Illicit Chinese small-scale mining in Ghana: Beyond institutional weakness? Sustainability. 2019;**11**(21):5943

[78] Snapir B, Simms DM, Waine TW. Mapping the expansion of galamsey gold mines in the cocoa growing area of Ghana using optical remote sensing. International Journal of Applied Earth Observation and Geo-Information. 2017;**58**:225-233

[79] Fahrig L. Effects of habitat fragmentation on biodiversity. Annual Review of Ecology, Evolution and Systematics. 2003;**34**:487-515

[80] Vanbergen A, Heard MS, Breeze TD, Potts SG, Hanley N. Status and Value of Pollinators and Pollination Services – A Report to the Department of Environment Fisheries and Rural Affairs (DEFRA). UK Centre for Ecology and Hydrology. UK: University of Stirling; 2014

[81] Ollerton J, Winfree R, Tarrant S. How many flowering plants are pollinated by animals? Oikos. 2011;**120**(3):321-326

[82] Blaser WJ, Oppong J, Hart SP, Landolt J, Yeboah E, Six J. Climate-smart sustainable agriculture in low-to-intermediate shade agroforests. Nature Sustainability. 2018;**1**(5):234-239. DOI: 10.1038/s41893-018-0062-8

[83] Jezeer RE, Verweij PA, Santos MJ, Boot RGA. Shaded coffee and cocoa – Double dividend for biodiversity and small-scale farmers. Ecological Economics. 2017;**140**:136-145. DOI: 10.1016/j.ecolecon.2017.04.019

[84] Toledo-Hernández M, Tscharntke T, Tjoa A, Anshary A, Cyio B, Wanger TC. Landscape and farm-level management for conservation of potential pollinators in Indonesian cocoa agroforests. Biological Conservation. 2021;**257**(109106):1-8. DOI: 10.1016/j.biocon.2021.109106

[85] Winfree R, Gross BJ, Kremen C. Valuing pollination services to agriculture. Ecological Economics. 2011;**71**:80-88

[86] Hadley AS, Betts MG. The effects of landscape fragmentation on pollination dynamics: Absence of evidence not evidence of absence. Biological Reviews. 2012;**87**:526-544

[87] Bommarco R, Marini L, Vaissiere BE. Insect pollination enhances seed yield, quality, and market value in oilseed rape. Oecologia. 2012;**169**:1025-1032

[88] Kennedy CM, Lonsdorf E, Neel MC, Williams NM, Ricketts TH, Winfree R, et al. A global quantitative synthesis of local and landscape effects on wild bee pollinators in agroecosystems. Ecology Letters. 2013;**16**:584-599

[89] Fischer J, Lindenmayer DB. Landscape modification and habitat fragmentation: A synthesis. Global Ecology and Biogeography. 2007;**16**(3):265-280

[90] Potts SG, Biesmeijer JC, Kremen C, Neumann P, Schweiger O, Kunin WE. Global pollinator declines: Trends, impacts and drivers. Trends in Ecology and Evolution. 2010;**25**(6):345-353

[91] Gyasi EA. Study of Climate Change Vulnerability and Adaptation Assessment Relative to Land Management in Ghana. Accra: Reports of the members of the study team; 2006. p. 91

[92] College BC. A guide to cocoa cultivation. In: A Hand Book for Practical Work. Ghana: Bunso Cocoa College; 1987. p. 35

[93] IPBES. Summary for Policymakers of the Global Assessment Report on Biodiversity and Ecosystem Services of the Intergovernmental Science-Policy Platform on Biodiversity and Ecosystem Services. Bonn, Germany: IPBES Secretariat; 2019

Section 4

Cocoa Cultivation: Role in Peace Building

Chapter 10

Cacao: A Path of Everyday Resistance and the Pursuit of Peace

Julian Villa-Turek Arbelaez

Abstract

Cacao crops are not only a form of subsistence and agricultural project but also connect the way of life and resistance in conflict scenarios. San José de Apartadó, Colombia, and its local population have suffered from the consequences of long-term armed conflict, which left hundreds of victims of forced displacement and disappeared, leaving a territory disconnected from its population and its interactions with agriculture and peace. The 2016 Peace Agreement between the FARC-EP and the Colombian government reopened a history of resistance among peasants, who have cultivated their lands to live and build peace by recognizing patterns of violence in the search for missing persons. Today, the armed conflict has not ended; there is a repeated presence of other armed groups in Urabá, a factor that involves the possibility for local populations to live in peace. Favorably, the institutions have begun to take action to continue with the efforts of the Peace Agreement. The creation of the Search Unit for Missing Persons (UBPD) has helped the families to continue searching for the disappeared and recognize the ways of life and practices of the territory, where cacao crops are a central form of life and social organization.

Keywords: cacao production, everyday resistance, search for missing people, peace research, human rights

1. Introduction

Colombia and various regions have exemplified how agriculture, particularly cacao crops, can foster peace and transform violence in areas affected by armed conflicts. In these regions, agriculture, connected to rural areas and land for production, has been a recurring source of territorial disputes involving various legal and illegal economies. San José de Apartadó, located in the northwest region of Urabá, is a referent of social leadership and resistance to powerful ways of violence deployed by armed actors and civilian groups. The cacao crops cultivated in this region have the characteristics of ensuing the central economy of thousands of families and, therefore, a form of resistance and strength to continue searching for their lost family members and relatives, aiming to reduce the impacts of a common framework of the Colombian armed conflict: the enforced disappearance. Special acknowledgment should be given to the social organization CACAOVIVE in San José de Apartadó, with whom I had the invaluable opportunity to conduct my bachelor's degree thesis in political science and emphasize peace research and conflict resolution in the region as a case study in 2020 and 2021. Through this research, I had the privilege of meeting various social leaders who dedicate themselves tirelessly to the pursuit of peace in the region[1] [1].

The enforced disappearance of civilians in the conflict and in San José de Apartadó was common during the armed conflict. Its use was aimed to eliminate the others, taking their humanity away from them, and ruining prebuilt social systems based on the relationships in communities with its territories and, therefore, their ways of life, practices, and knowledge. For the National Centre for Historical Memory (CNMH), more than 60,000 people have disappeared between 1970 and 2015, contrasting with the Registry of Victims (RUV), a State Registry to know how many people have the status of victims. They may need assistance for reparation and Truth—who aims that between 1978 and 2020, at least 178.000 were victims of this type of violence. Although these statistics may be suffering sub-registry omissions due to labels of murders and kidnappings—related forms of violence—they show how big numbers are and the challenges to finding them inside the armed conflict and the still presence of armed groups in the region.

The use of violence, including enforced disappearance, has been a prominent and widely acknowledged issue, which was extensively discussed during the peace negotiations held in La Habana, Cuba. These negotiations ultimately led to signing the 2016 Peace Agreements between the Colombian State and the Revolutionary Armed Forces of Colombia (FARC-EP) guerrilla group. The primary objective of these agreements was to address historical conflicts rooted in territorial and social issues. Furthermore, one of the highlights has been the creation of The Colombian Integrated System of Truth, Justice, Reparation, and Non-Repetition (SIVJRNR in Spanish), in which the victims are central key players in peacebuilding alongside new policy frameworks and bodies such as the Truth Commission (*CEV* in Spanish), the Special Jurisdiction for Peace (*JEP* in Spanish) and the Search Unit for Missing Persons (*UBPD* in Spanish). The last have developed different strategies to strengthen confidence between individuals, collective actors, and the State to search, find, prospect, and identify missed persons, victims of enforced disappearance during the armed conflict.

With pedagogical tools, the Unit started approaching different collective and individual actors. Many social organizations stated social objectives in specific territories with differential characteristics and armed conflict impacts, just as in San José de Apartadó and the Urabá region in Colombia. In this region, the Unit implemented the Creative Knowledge Circles (*Círculos de Saberes Creativos* in Spanish), a dissimilar methodology to the judicial system. This approach aimed to listen to the victims in a horizontal relationship, acknowledging the knowledge and practices they have.

This research endeavors to establish a connection between social leadership, which centers around developing the local economy through cacao crops, and peacebuilding over the search for missing people due to the armed conflict. A particular emphasis is placed on the utilization of Creative Knowledge-Circles to strengthen the outcomes of dialog between different stakeholders. The objective is to contribute to the search for missing persons while concurrently promoting the establishment of sustainable cacao-based agriculture as a more equitable approach to foster local economies.

[1] Undergraduate thesis in Political Science with emphasis on Conflict Resolution and Peace Research at the Pontificia Universidad Javeriana, Bogotá, Colombia.

2. Theoretical framework: Everyday resistance and searching for missing people

The exercise of social leadership in a place fully affected by violence implies the use of categorical variables to analyze how the search for missing people and the use of cacao crops allow a new understanding with the case study based on qualitative methodologies as a guiding example. First, the resistance to violence can be addressed with an everyday resistance approach. Second, the search for missing people constitutes a category of analysis that can be connected to different types of organizations. Third, the connection between both categories could influence the agricultural choices of a specific population in a territory, as the relationship with State bodies and institutions.

Resistance to violence has been studied in social sciences as a tool to understand everyday violence management. Scott [2] defined everyday resistance as a process in which resist means to start organizing from small actions and a discursive tool over a specific time, space, and with different social relationships. Since violence can be considered a state of power [2–4], society can decide to change that statement of violence in everyday contexts. Therefore, a historically related power practice can be reactive to a homogeneous and noncontingent context, the reason behind its intersectional dimension to consider different types of power and changes [4]. How resistance can arise after violence disclaimed collective visions to invisible ones, the same that starts to change since the first invisible and individual stage that is open to establishing a new political subject.

Purposes of resistance distinguish between purposes of violence since it is not singular and unique in its patterns and identities. For authors like Butler [5] and Mbembe [6], violence follows patterns of affecting individuals and collectives due to an interest in affecting the humanity of others and making invisible race, gender, and cultural differences. The body is at the mercy of social and environmental modes that allow the use of war, in which death is justified and approved toward some subjects. Therefore, the use of violence and resistance to it allows the creation of new spheres of recognition and concession to other existing epistemologies that have been disrupted by different uses of power. This recognition of everyday resistance to contexts marked by armed conflicts allows the recognition of specific uses of violence in which social efforts to resist will be held.

Since forced disappearance has been shared in some armed conflicts, the theory helps the Colombian case analysis. This kind of violence consists of repression methods that break social and individual senses of identity and practices [7]. According to the classical framework created by Galtung [8], this violence can be defined as direct violence (visible) and addressed within a structure with cultural conditions to legitime the use of violence [9]. But the effect of the forced disappearance affects not just the missed ones and victims but also the ones who stay and start looking for them, victims too. Thus, as Casado [10] expresses, the interest in searching for the lost defines the search as a bioprocess of who is being searched for and who is searching for them.

A subject who decides to search has become a political actor with a differentiated identity. Since it is also a victim of forced disappearance and has faced the process of invisiblization of the pattern of violence, the remembering process deeply suffers and puts the searcher in a complex process of recognizing their identity and practices [9, 11]. Anonymity begins to fade from the moment an individual begins to search, a process that continually has challenges, as organizing a search group is a complicated second step due to the breakdown of social networks and the impacts on identity left by violence and enforced disappearance.

The missed ones have affected left a space of uncertainty in the daily lives of their families, neighbors, and known people. Therefore, authors like Delacroix [12], analyzing the case of Perú and Robledo [13] in Mexico, have stated the significant impacts on everyday life with an absence of meaning that leads to a new performance of stories that connect with others who are living the same on rebuilding step process. Since the impacts left a catastrophe on social networks [13], it connects a reduced interpersonal network that becomes a powerful tool to remember the missing to act alongside and on them [12]. These purposes and processes may be accompanied by institutional efforts [10]. Then, the establishment of new public policies toward the search for missed people states new practices to manage the absence of institutions or stigmatization and revictimization of the victims. That guides the collective subject that Jelin [14] mentions, with an agency built on interpersonal relations that can work or demand institutions to start, continue, or improve the search for missing people. All these efforts need to be recognized as an everyday process as they improve the achievements to change structural violence and seek conflict resolutions and peacebuilding.

3. Context over the study case

3.1 Cacao crops in Colombia: A more visible economy

In Colombia, at least three cacao crops can be planted in the country. Under official statistics [15], more than 65.000 families live in 422 municipalities and 27 departments where this agriculture has been deployed. In terms of production, more than 188.000 Ha in 2020, where 64.000 tons were produced. Also, each producer produces at least 3 Ha of cacao crops in the median. The regions with the most cacao production are those that the armed conflict has deeply impacted. In the last decade, cacao production expansion mainly took place in remote, low-connected, and low-density areas such as the Choco, Amazon foothills, and Magdalena Medio. Notably, these areas have also been highly impacted by the internal armed conflict [16]. In this sense, the entry of more extractivist multinationals represents a great vital and territorial threat to the entire country and especially to the Gulf of Urabá region [17]. Therefore, cacao yield had a spatial and temporal trend characterized by an increase in regions that were previously impacted by the conflict, such as Urabá, creating instability while impacting smallholders' likelihood to receive training, credits, and their capacity to produce permanent crops (see **Figures 1–3**) [16].

3.2 San José de Apartadó: A history of violence and hope

San José de Apartadó has been living a story of violence and resistance. Since the 1970s, civilian organization based on the creation of banana agrarian unions formed in the 1960s, and the guerrillas' support increased violence with forced displacement, the first massacres of civilians, and the forced disappearance of people during the armed conflict with the Colombian army [18]. Years later, a systemic way of repression was deployed to dismantle and affect a region where the political party of Unión Patriótica (UP) – part of the Peace Treaty Agreements between the State and the guerrilla of the FARC-EP in 1984 – had enormous support since they used to guide a new political agenda based on social leaderships. Although it had high pikes in the decade of 1980, the next decade came up with new enforcement in violence levels and the entry of paramilitary groups to the region [19, 20]. The Autodefensas Unidas

DOI: http://dx.doi.org/10.5772/intechopen.112999

Figure 1.
Location of Urabá region in the Department of Antioquia and San José de Apartadó. Source: Movimiento regional por la Tierra [17].

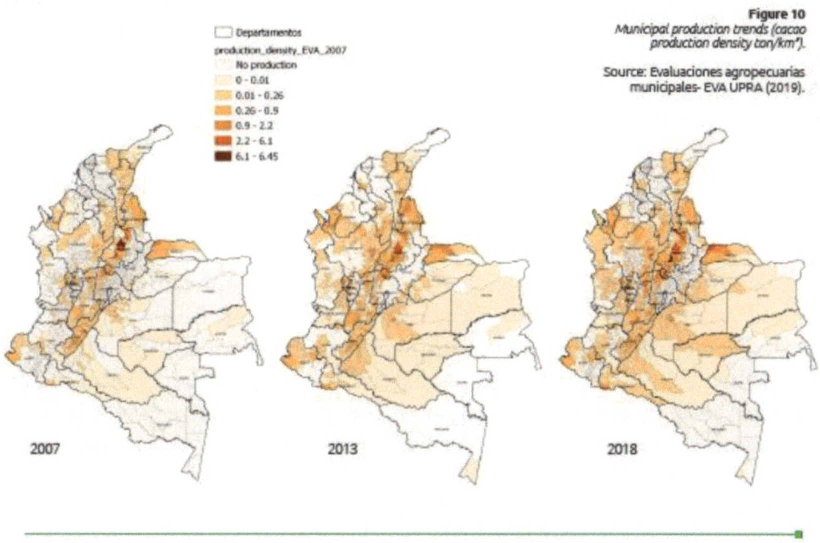

Figure 2.
Municipal production trends (cacao production). Source: Grow Colombia [16].

de Colombia (AUC) paramilitary group, one of the biggest ones, perpetrated more massacres and was a crucial factor in the increase of missing people due to the armed violence and the social control deployed by armed groups. From 1994 onwards, just in San José de Apartadó, the number of cases of disappearance increased without a

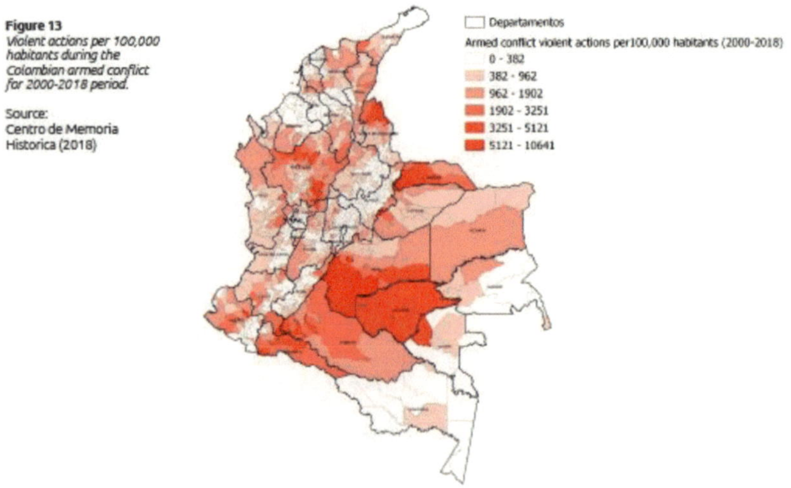

Figure 3.
Violent actions per 100.000 habitants during the Colombian armed conflict for the 2000–2018 period. Source: Centro Nacional de Memoria Histórica (CNMH) (2018) cited in Grow Colombia [16].

notable decrease until 2004, with more than 100 cases per year. The municipality remains in the first five cities in the Department of Antioquia with more cases of enforced disappearance.

As a form of resistance to violence, in 1997, a group of more than 300 families created the Peace Community of San José de Apartadó (CPSJA). It became a neutral place where no armed group could enter its territory. Therefore, they had to deal with all armed groups at different times, facing a continuum of violence and stigmatization of all armed groups, including guerrillas, paramilitaries, and national armed forces [21–24]. After the 2005 massacre against the Peace Community, the demobilization of the AUC dismantled the Paramilitary Central Bloc in Urabá (BEC) as a result of negotiations between the AUC and the central government that ended in the Santa Fé de Ralito Agreements of 2003. Although an agreement was reached, the paramilitary groups did not disappear and were transformed into more decentralized armed groups, as were the criminal gangs [19, 20]. A decrease in violence was apparent and visible but far from ending. However, the transformation of the armed conflict led to new armed groups that remain predominant to the present day as the National Liberation Army (ELN) and the Gaitanista Self-Defense Forces of Colombia (AGC), which reflects the aggravated context against the settlers and the general fragmentation of trust with the Colombian State [17, 23]. However, the Peace Agreements of 2016 opened a new window of opportunity to build peace on the territorial level alongside civil society. The creation of the SIVJRNR influenced the creation of the Search Unit for Missing Persons (UBPD). This humanitarian and extrajudicial entity has operated autonomously and independently from the traditional judicial bodies.

In the case of agriculture and cacao production, most producers are individuals who, together with cooperatives, cultivate the land with cacao crops and other plantations (for their own use, e.g., corn, banana, avocado and mandioca/cassava). In the last decades, there was a change from banana to cacao production as the primary plantation since the first one raised price problems, and a new market of exportation created the input for the organizations to think and design new strategies [17]. Thanks to Peace Accords, the figure of the cooperatives became an excellent opportunity to transform social leadership, not only in the organizational and political view but also in the economic model. It influenced a new recognition by the State and the territory and its legitimate practices, such as cocoa cultivation and production.

3.3 Creative knowledge circles strategy

Implementing different policies after the Peace Agreements of 2016 has changed how the State searches for missing people due to the armed conflict. Under the new policy frameworks, the primary mission is to fulfill victims' rights to reparation, access to truth and justice, and guarantees for non-repetition. These agendas serve as guiding principles for the institutions established because of the 2016 Peace Agreements and build upon the foundation laid by the Law 1448 of 2011, which aimed to recognize the armed conflict and address reparations to the victims. Then, the role of the UBPD has been searching for people who may be victims of enforced disappearance and to differentiate its actions from those carried out by existing institutions such as the Attorney General's Office. Its approach consists of five steps in searching for missing people: search, location, recovery, identification, and dignified return. These steps follow the interest of working with participation, information, and prospecting since the need to add voices and stories of victims is crucial to have a human searching and bring all the experiences of professionals to prospect, recover, and identify identities, aiming to a humanitarian and extrajudicial perspectives [1].

The creative knowledge circles mentioned above can be defined as an innovative way involving the active participation of victims and groups who have experienced disappearances. With a safe and inclusive space, all individuals can share their stories and connect the experiences of violence and resistance (on an individual or collective level) and the ongoing efforts to locate missing persons. In those spaces, the UBPD plays a crucial role in facilitating the participation of all involved parties and ensuring their voices are heard. The institution guarantees its support by collecting and managing information to initiate the process of locating potential sites of violent incidents or places where information on missing individuals may be available. Furthermore, the UBPD utilizes genetic identification techniques to identify living persons separated from their families due to violence. This whole process is therefore an effort that over time, requires the gradual establishment of trust and rapport between the individuals. As trust is built, progress can be made in gathering and assessing the information related to the missing individuals and the violent episodes they experienced.

4. Analysis: How cacao crops can be a good tool to build peace and help the UNBPD efforts to find missing people in San José de Apartadó

The decision of communities in San José de Apartadó to focus on cacao cultivation as their primary agricultural product stems from their aspiration for better and more equitable local economies, intending to resist the violence prevalent in the region. The cultivation of cacao has emerged as one of the most significant economic activities for local families, influencing their resistance against not only the armed conflict and the recruitment of young people (due to a lack of preventative policies) but also the cultivation of illicit crops such as marijuana and coca, which directly undermines community peacebuilding efforts.

The social leaders of the CACAOVIVE organization assert that the fertile land in the region can support the cultivation of various crops. However, cacao remains the predominant choice due to several factors. First, cacao is a resilient crop that does not easily succumb to damage or diseases. It is also relatively cost-effective to cultivate, even without fair pricing mechanisms in the market [1, 16, 17].

Regarding pricing, the leaders argue that multinational companies often offer a reasonable and fair price for the crop during the cacao-growing season. However, this is not the case during off-season periods when the farmers' demand and prices are less favorable. Despite this challenge, the communities continue to prioritize cacao cultivation as it represents their primary source of income and a means to sustain their livelihoods amidst the prevailing socioeconomic circumstances. By focusing on cacao cultivation, the San José de Apartadó communities aim to create economic opportunities that provide for their immediate needs and contribute to their resilience and resistance against the violence and illicit economies that undermine peacebuilding efforts in the region.

After resisting decades of violence, the communities decided to organize themselves. Within a story of colonization over the territory and enforced displacement from armed actors and multinationals of banana cultivation since the 1970s, cultivating the land became a way of resistance [18, 21]. Since the physical and psychosocial problems have been extreme, the interest in sitting with the community and deciding on creating social organizations and cooperatives implies the first step in everyday resistance. They could reveal positions and interests to recognize the impacts of the armed conflict and work to transform them [3]. Then, the everyday resistance started from the individual's perspective but had deep transformations to recognize marginalized people [6] and build toward cultural differences that connect to the territory and the cultivation of cacao [5]. These efforts also connect with an institutional framework after creating cooperatives and the interest to work alongside public institutions on common challenges.

Therefore, the efforts of cooperatives and social organizations may include working alongside institutions and new bodies such as the UBPD. Implementing the creative circles' strategy has been crucial in this regard due to the recognition of the history of resistance of families searching for missing persons without legal assistance. The answer to understanding the change of collaborative work among actors is related to building trust among the parties, including members of the same community and territory. After a long and sustained stigmatization by most state institutions toward victims throughout the country, coupled with a lack of policies that address the needs of this population, the entry of institutions such as the UBDP has transformed how the search is understood as humane with the requirement of having all the knowledge and practices of civil society, becoming a horizontal hierarchy when it comes to dialog and as a starting point for locating them.

In the Circles of Knowledge, the story of production of cacao has played a crucial role, providing insights that guide families in San José de Apartadó. This implies that families of missing persons participate alongside individuals who are not involved in the search. Consequently, this connection aligns with the bioprocess defined by Casado [10], where the shared interest in searching for missing persons brings together both victims and non-victims, affected by both structural and cultural violence [8]. Therefore, the Circles seek to promote the exchange of knowledge and practices among participants with participation methodological tools. An open exchange with other actors may arise when there is a transparent image of the resistance with discursive rhetoric [3]. Through a transformative know-how process [25], the fact that the UBPD promotes these spaces strengthens the recognition based on the community's differences from other actors and the rich set of practices they deploy, including the search and the cultivation of cacao.

Creating new stories to transform social networks has been an ongoing and arduous effort undertaken by social organizations. Since the enforced missing can be defined as a catastrophe [13], the absence of the people alters how processes may be conducted in the community. The organizations have been working on different spheres over time, dividing tasks and activities related to the booster of conditions to cultivate the land with cacao crops, and also addressing human rights matters by communicating to authorities and more NGOs about risk levels that social leaders and the civil population face due to the presence of armed groups and the stigmatization that may arise between judicial institutions, the national army and the civil organizations. That creates a collective subject that is organized and ready to continue without institutional support [14] but with an everyday resistance approach [4]. The claim for a total memory without more blood has been a statement for civil organizations, such as the cooperatives and local participation bodies.

Moreover, it is crucial to consider the perspective of the UBPD workers, who recognize the importance of adopting an active listening approach while participating in spaces like Creative Circles. Through this exchange of knowledge, two aspects arise. First, understanding the complex social context is characterized by visibly armed conflicts that persist. Second, they become aware of communities' relentless efforts in searching for missing people and their organized approach, which connects the agricultural way of life and the decision to cultivate cacao crops as a way of resistance means of resistance against an unstable market and the lack of supportive programs from the State for small-scale producers to improve their crop systems. After the paramount cause of searching for the missing people, they unveil the strength and knowledge within these communities, who are determined to persist and resist through their way of life. Therefore, cacao production is a sign of resistance to unequal markets in the territory, systemic violence, and a State that has failed to provide adequate assistance for developing local economies and supporting peacebuilding.

Consequently, the development of pedagogical tools around the Circles of Knowledge has recognized the power of resistance through a territorial and communitarian organization. A new type of organization improves the recognition level by the State and its bodies of the historical violence that has affected the territory and population, as is the case of San José de Apartadó and the Urabá region in Colombia. Second, it has stated that social leaders are developing, with their possibilities and available resources, an integrative system of action that is co-guided by human rights advocacy and the establishment of sustainable economies that allow them to create a community atmosphere through more spaces of participation.

5. Conclusions

It has been seen around this research that searching for missing people in San José de Apartadó is an organizational process that has been active in the last decades after experiencing the profound impacts of violence left in the territory and the communities. Despite the lack of public policies that could create new public management approaches to create better cultivation facilities and spaces for education, the communities organized through social organizations and cooperatives have led the image of a reborn resistance to the reconstruction of the social fabric. The entrance of the UBPD has been a clear signal of how State bodies can reconnect with citizens after a long-stigmatized relationship due to the armed conflict. Their entrance means a reconstruction of confidence and allows the institutions to support already designed community tasks to search for the missing and establish local economies based on cacao production.

Nonetheless, the challenges to building peace remain in Urabá and other regions of Colombia. The 2016 Peace Agreements do not mean the violence has been decreasing, nor by the new government of Gustavo Petro, a progressist who is willing to negotiate with all the armed groups, creating the so-named "Total Peace", a way full of challenges and step backs, a situation that shows how the civil organizations remain resisting to live in peace on its territories. In general, the resistance, shown as an everyday task in connection to the cultivation of the land, will remain a key factor of peacebuilding from a bottom-up approach. The replication of development models that maintain the violence and armed conflict characteristics may create new conflicts, which is the reason behind the importance of spaces like the UBPD ones, in which the exchange of knowledge can create a better understanding of the territory and the possibilities of cultivating cacao crops as a territorial peace tool.

The research on cacao crops and everyday resistance has significant potential. It can significantly benefit from interdisciplinary approaches since the possibilities are enormous, and the interest in having more sustainable economies, energy transition platforms, and the protection of the environment can be a clear advantage. Bringing in perspectives from diverse disciplines, such as agriculture, environmental sciences, economics, social sciences, and sustainability, would be relevant for efforts to move toward more sustainable crops and production, ensuring equitable access to assistance and markets at the national and international level. Listening to the leaders of San José de Apartadó has been a crucial experience for redesigning policies and creating better results in peacebuilding and violence transformation processes.

Acknowledgements

I want to sincerely thank the following individuals and organizations for their invaluable contributions and stories of resistance: Néstor, Luis, Leonel, Fredy, and all the people from San José de Apartadó. I would also like to extend my gratitude to the whole CACAOVIVE organization for its leadership and defense of human rights. I am deeply grateful to my supervisor, Juan Daniel Cruz, for their guidance and friendship. Lastly, I would like to express my gratitude to my dear friends and research teammates, Juan Esteban Uribe and Sebastián Osorio, who, together with Juan Daniel, have supported and collaborated to our journey in peace research and decolonial peace.

Cooperativa CACAOVIVE: https://www.facebook.com/profile.php?id=100081035856701&locale=hi_IN

Author details

Julian Villa-Turek Arbelaez
Hertie School, the University of Governance in Berlin, Germany

*Address all correspondence to: julianvillat@hotmail.com

References

[1] Villa-Turek Arbeláez J. Vivir Para Buscarlos: Case Study of Cacao Vive in San José de Apartadó and its Search for Missing Loved Ones as a Nonviolent Strategy [Master's thesis]. Bogotá, Colombia: Pontificia Universidad Javeriana; 2021. Available from: https://repository.javeriana.edu.co/handle/10554/54509

[2] Scott JC. Everyday forms of resistance. In: Peasant Economy, Culture, and Politics of European Russia, 1800-1921. Princeton University Press; 1989. p. 236

[3] Vinthagen S, Johansson A. "Everyday resistance": Exploration of a concept and its theories. Resistance Studies Magazine. 2013;**2013**:1

[4] Martínez Bernal DR. La Noviolencia en los Nasa, del norte del Cauca: relaciones entre la teoría y la experiencia específica. Polis (Santiago). 2016;**15**(43):159-180

[5] Butler J. Marcos de guerra: las vidas lloradas. Barcelona, España: Paidós básica; 2009

[6] Mbembe A. Necropolítica seguido de Sobre el gobierno privado indirecto. In: Archambault EF, editor. Editorial Melusina; 2011

[7] Faúndez Abarca X, Azcárraga Gatica B, Benavente Morales C, Cárdenas CM. La Desaparición Forzada de Personas a Cuarenta Años del Golpe de Estado en Chile: un Acercamiento a la Dimensión Familiar. Revista Colombiana de Psicología. 2018;**27**(1):85-103

[8] Galtung J. Violencia, guerra y su impacto. Sobre los efectos visibles e invisibles de la violencia. 2004

[9] Hernández Medina AC. Desaparición forzada de personas como estrategia de violencia Un estudio de la desaparición forzada de 43 campesinos por parte del grupo paramilitar "Los Tangueros" en el Corregimiento de Pueblo Bello (Turbo, Antioquia). 2016

[10] Casado ND. Las "prácticas del retorno" en la desaparición de personas: la materialización más allá de la ausencia. Sociology & Technoscience/Sociología y Tecnociencia. 2019;**9**(2):99-114

[11] Schulz C, Salazar M. La búsqueda de la verdad: una necesidad para tejer el futuro. La búsqueda de personas desaparecidas desde una perspectiva de los derechos humanos. Cotidiano - Revista de la Realidad Mexicana. 2018;**33**(209):103-109

[12] La DD. Hacia una antropología de la vida póstuma de los desparecidos en el Perú. Íconos.Revista de Ciencias Sociales. 2020;**67**:61-74

[13] Robledo SC. Genealogía e historia no resuelta de la desaparición forzada en México. Íconos. 2016;**20**(55):93

[14] Jelin E. Los derechos humanos y la memoria de la violencia política y la represión: la construcción de un campo nuevo en las ciencias sociales. Cuadernos del Ides. 2003;**2**:3-27

[15] Ministerio de Agricultura. CADENA DE CACAO. Dirección de Cadenas Agrícolas y Forestales. 2021

[16] GROW Colombia Project. Socioeconomic Programme. Perspectives on a Bio-Economy Development Path for Colombia. Cacao farming for peace building and rural development. Report 4. 2022

[17] Movimiento Regional por la Tierra. Comunidad de Paz de San José de

Apartadó: Ruptura con el Estado como resistencia a la guerra. 2017

[18] Mazzoldi Díaz G, Huertas Montoya RA, Cuesta Astroz I, Tobo Caviedes PA. Territorio, seguridad y violencias basadas en género en Apartadó. Reyes Le Paliscot E, Franco M, editors. Bogotá D.C., Colombia: Fundación Ideas para la Paz - FIP; 2019 p. 1-82. (Fundación Ideas para la Paz - FIP; vol. 2)

[19] Popular CC de I y E. Panorama de los derechos humanos y violencia política en Colombia. Bogotá, DC, Colombia: CINEP/PP Centro de Investigación y Educación Popular; 2005

[20] Giraldo MJ, Fusil o Toga. Tuga y Fusil. Colombia: Editorial Códice Ltda; 2010

[21] Aparicio Cuervo JR. El retorno a Mulatos y la Comunidad de Paz de San José de Apartadó: contingencias y momentos de ruptura. 2015;**21**:73-95. Available from: http://www.scielo.org.co/scielo.php?script=sci_arttext&pid=S1900-54072015000100004&lng=en&tlng=en

[22] Muñoz Lopera JM. La paz bajo la sombra indeleble de la guerra: una experiencia de construcción de ciudadanía en la comunidad de paz de San José de Apartadó (Antioquia, Colombia). 2008;**22**(39):159-184

[23] Courtheyn C. Comunidad de paz: una paz 'otra' en San José de Apartadó-Colombia. Polisemia. Jul 2016;**12**(22):55-72

[24] San ZR, José de Apartadó, Comunidad de Paz: La libertad como instinto de sobrevivencia. 2011. Available from: https://www.alainet.org/es/articulo/146912

[25] Sandoval EA. Educación para la paz integral Memoria, Interculturalidad y decolonialidad. 2016